The Universe

Books by John Brockman

AS AUTHOR:

By the Late John Brockman
37
Afterwords
The Third Culture: Beyond the Scientific Revolution
Digerati

AS EDITOR:

About Bateson
Speculations
Doing Science
Ways of Knowing
Creativity
The Greatest Inventions of the Past 2,000 Years
The Next Fifty Years
The New Humanists
Curious Minds
What We Believe but Cannot Prove
My Einstein
Intelligent Thought
What Is Your Dangerous Idea?
What Are You Optimistic About?
What Have You Changed Your Mind About?
This Will Change Everything
Is the Internet Changing the Way You Think?
The Mind
Culture
This Will Make You Smarter
This Explains Everything
What Should We Be Worried About?
Thinking

AS COEDITOR:

How Things Are (with Katinka Matson)

THE
Uni-
verse

Leading Scientists Explore
the Origin, Mysteries, and
Future of the Cosmos

EDITED BY

John Brockman

HARPER  PERENNIAL

NEW YORK • LONDON • TORONTO • SYDNEY • NEW DELHI • AUCKLAND

FIRST EDITION

Designed by Michael Correy

Library of Congress Cataloging-in-Publication Data has been applied for.

ISBN 978-0-06-229608-5

14 15 16 17 18   OV/RRD   10 9 8 7 6 5 4 3 2 1

I wish to thank Peter Hubbard of HarperCollins for his encouragement. I am also indebted to my agent, Max Brockman, for his continued support of this project.

Contents

Introduction

In this, the fourth volume of The Best of *Edge* series, following *Mind, Culture,* and *Thinking*, we focus on ideas about the universe. We are pleased to present twenty-one pieces, original works from the online pages of *Edge.org*, which consist of interviews, commissioned essays, and transcribed talks, many of them accompanied online with streaming video.

Edge, at its core, consists of the scientists, artists, philosophers, technologists, and entrepreneurs at the center of today's intellectual, technological, and scientific landscape. Through its lectures, master classes, and annual dinners in California, London, Paris, and New York, *Edge* gathers together the "third-culture" scientific intellectuals and technology pioneers exploring the themes of the post-industrial age. These are the people who are rewriting our global culture.

And its website, *Edge.org*, is a conversation. The online *Edge. org* salon is a living document of millions of words that charts the conversation over the past eighteen years. It is available, gratis, to the general public.

Edge.org was launched in 1996 as the online version of the Reality Club, an informal gathering of intellectuals that met from 1981 to 1996 in Chinese restaurants, artists' lofts, the boardrooms of Rockefeller University and the New York Academy of Sciences, investment banking firms, ballrooms, museums, living rooms, and elsewhere. Though the venue is now in cyberspace, the spirit of the Reality Club lives on, in the lively back-and-forth conversations on hot-button ideas driving the discussion today. In the words of novelist Ian McEwan, a sometime contributor, *Edge.org* is "open-minded, free ranging, intellectually playful . . . an unadorned pleasure in curiosity, a collective expression of wonder at the living and inanimate world . . . an ongoing and thrilling colloquium."

This is science set out in the largely informal style of a conversation among peers—nontechnical, equationless, and colloquial, in the true spirit of the third culture, which I have described as consisting of "those scientists and other thinkers in the empirical world who, through their work and expository writing, are taking the place of the traditional intellectual in rendering visible the deeper meanings of our lives, redefining who and what we are."

For this volume—coming as it does in the wake of the recent stunning discovery of gravitational waves by the BICEP2 radio telescope at the South Pole, an apparent confirmation of the prime cosmological theory of inflation—we've assembled online contributions from some of *Edge's* best minds, most of them pioneering theoretical physicists and cosmologists. They provide a picture of cosmology as it has developed over the past three decades—a "golden age," in the words of MIT's Alan Guth, one of its leading practitioners.

This Golden Age of Cosmology has reached a high point—not just with the recent revelations at the South Pole but with the 2012 discovery, by the Large Hadron Collider at CERN, of the long-sought Higgs boson, whose field is thought to give mass to the elementary particles making up the universe.

We lead off, appropriately enough, with a 2001 talk by Guth, the father of inflationary theory. Then, at an *Edge* gathering at Eastover Farm in Connecticut a year later, Guth goes head to head with Paul Steinhardt, who presents his rival theory of a cyclic universe—a theory that the new data on gravitational waves may put paid to. Stay tuned.

Andrei Linde, the father of "eternal chaotic inflation," emphasizes the concepts of the multiverse and the anthropic principle that arose from it (" . . . different exponentially large parts of the universe may be very different from each other, and we live only in those parts where life as we know it is possible").

Lisa Randall and Neil Turok elaborate on the theory of branes, two-dimensional structures arising from string theory—and whose existence is central to the cyclic universe.

Sean Carroll ponders the mystery of "why our observable universe started out in a state of such pristine regularity and order."

Martin Rees, the U.K.'s Astronomer Royal, speculates on whether we are living in a simulation produced by a superintelligent hypercomputer.

Lee Smolin discusses the nature of time. Then he and Leonard Susskind (string theory's father) engage in a gentlemanly donnybrook over Smolin's theory of cosmological natural selection and the efficacy of the anthropic principle.

Brian Greene, Paul Steinhardt, and Einstein biographer Walter Isaacson speculate on how Einstein might view the theoretical physics of the 21st century, and Steinhardt and Greene come to gentlemanly blows over string theory.

In a calmer vein:

Science historian Peter Galison muses on the similarity, and fundamental dissimiliarity, between two contemporaries and giants of early 20th-century physics, Einstein and Poincaré; Arizona State University cosmologist Lawrence Krauss throws up his hands at the conundrum of dark energy; Carlo Rovelli (*Professeur de classe exceptionelle, Université de la Méditerranée*) recommends a willingness to return to basics; and Nobelist Frank Wilczek relishes a future devoted to "following up ideas in physics that I've had in the past that are reaching fruition."

Berkeley's Raphael Bousso, too, is an optimist. ("I think we're ready for Oprah, almost . . . [W]e're going to learn something about the really deep questions, about what the universe is like on the largest scales, how quantum gravity works in cosmology.")

Seth Lloyd, a quantum mechanical engineer, explains how the universe can, in a sense, program itself; Steven Strogatz sees cosmic implications in the synchronous flashing of crowds of fireflies; and Oxford physicist David Deutsch predicts that his "constructor theory" will eventually "provide a new mode of description of physical systems and laws of physics. It will also have new laws of its own, which will be deeper than the deepest existing theories such as quantum theory and relativity."

And finally, as a kind of envoi, the late Benoit Mandelbrot, nearing eighty, looks back on a long career devoted to fractal geometry and newly invigorated: "A recent, important turn in my life occurred when I realized that something I have long been stating in footnotes should be put on the marquee. . . . I'm particularly long-lived and continue to evolve even today. Above a multitude of specialized considerations, I see the bulk of my work as having been directed toward a single overarching goal: to develop a rigorous analysis for roughness. At long last, this theme has given powerful cohesion to my life."

A Golden Age of Cosmology it may be, but you will find plenty of roughness—of doubt and disagreement—here. In spite of its transcendent title, this collection is hardly the last word. In the months (and the years) ahead, as the Large Hadron Collider pours out more data about the microworld and powerful telescopes and satellites continue to confirm—or, who knows, cast fresh doubt on—our leading theories of the macroworld, the arguments will surely continue.

May the conversation flourish!

John Brockman
Editor and Publisher
Edge.org

The Universe

1
A Golden Age of Cosmology

Alan Guth

Father of the inflationary theory of the Universe and Victor F. Weisskopf Professor of Physics at MIT; inaugural winner, Milner Foundation Fundamental Physics Prize; author, *The Inflationary Universe: The Quest for a New Theory of Cosmic Origins*

It's often said—and I believe this saying was started by the late David Schramm—that today we are in a golden age of cosmology. That's really true. Cosmology at this present time is undergoing a transition from being a bunch of speculations to being a genuine branch of hard science, where theories can be developed and tested against precise observations. One of the most interesting areas of this is the prediction of the fluctuations, the nonuniformities, in the cosmic background radiation, an area I've been heavily involved in. We think of this radiation as being the afterglow of the heat of the Big Bang. One of the remarkable features of the radiation is that it's uniform in all directions, to an accuracy of about 1 part in 100,000, after you subtract the term that's related to the motion of the Earth through the background radiation.

I've been heavily involved in a theory called the inflationary universe, which seems to be our best explanation for this uniformity. The uniformity is hard to understand. You might think initially that maybe the uniformity could be explained by the same principles of physics that cause a hot slice of pizza to get cold when you take it out of the oven; things tend to come to a uniform temperature. But once the equations of cosmology were worked out so that one could calculate how fast the universe was expanding at any given time, physicists were able to calculate how much time there was for this uniformity to set in.

They found that in order for the universe to have become uniform fast enough to account for the uniformity we see in the cosmic background radiation, information would have to have been transferred at approximately a hundred times the speed of light. But according to all our theories of physics, nothing can travel faster than light, so there's no way this could have happened. So the classical version of the Big Bang theory had to simply start out by assuming that the universe was homogeneous—completely uniform—from the very beginning.

The inflationary universe theory is an add-on to the standard Big Bang theory, and basically what it adds on is a description of what drove the universe into expansion in the first place. In the classic version of the Big Bang theory, that expansion was put in as part of the initial assumptions, so there's no explanation for it whatsoever. The classical Big Bang theory was never really a theory of a bang; it was really a theory about the aftermath of a bang. Inflation provides a possible answer to the question of what made the universe bang, and now it looks like it's almost certainly the right answer.

Inflationary theory takes advantage of results from modern particle physics, which predicts that at very high energies there should exist peculiar kinds of substances which actually turn gravity on its head and produce repulsive gravitational forces. The inflationary explanation is the idea that the early universe contains at least a patch of this peculiar substance. It turns out that all you need is a patch; it can actually be more than a billion times smaller than a proton. But once such a patch exists, its own gravitational repulsion causes it to grow rapidly, becoming large enough to encompass the entire observed universe.

The inflationary theory gives a simple explanation for the uniformity of the observed universe, because in the inflationary model the universe starts out incredibly tiny. There was plenty of time for such a tiny region to reach a uniform temperature and uniform density, by the same mechanisms through which the air in a room reaches a uniform density throughout the room. And if you isolate a room and let it sit long enough, it will reach a uniform temperature as well. For the tiny universe with which the inflationary model begins, there's enough time in the early history of the universe for these mechanisms to work, causing the universe to become almost perfectly uniform. Then inflation takes over and magnifies this tiny region to become large enough to encompass the entire universe, maintaining this uniformity as the expansion takes place.

For a while, when the theory was first developed, we were worried that we would get too much uniformity. One of the amazing features of the universe is how uniform it is, but it's still by no means completely uniform. We have galaxies and stars and clusters and all kinds of complicated structure in the universe that needs to be explained. If the universe started out completely uniform, it would just remain completely uniform, as there would be nothing to cause matter to collect here or there or any particular place.

I believe Stephen Hawking was the first person to suggest what we now think is the answer to this riddle. He pointed out— although his first calculations were inaccurate—that quantum effects could come to our rescue. The real world is not described by classical physics, and even though this was very high-brow physics, we were in fact describing things completely classically, with deterministic equations. The real world, according to what we understand about physics, is described quantum mechanically, which means, deep down, that everything has to be described in terms of probabilities.

The "classical" world we perceive, in which every object has a definite position and moves in a deterministic way, is really just the average of the different possibilities the full quantum theory would predict. If you apply that notion here, it's at least qualitatively clear from the beginning that it gets us in the direction we want to go. It means that the uniform density, which our classical equations were predicting, would really be just the average of the quantum mechanical densities, which would have a range of values that could differ from one place to another. The quantum mechanical uncertainty would make the density of the early universe a little bit higher in some places, and in other places it would be a little bit lower.

So, at the end of inflation, we expect to have ripples on top of an almost uniform density of matter. It's possible to actually calculate these ripples. I should confess that we don't yet know enough about the particle physics to actually predict the amplitude of these ripples, the intensity of the ripples, but what we can calculate is the way in which the intensity depends on the wavelength of the ripples. That is, there are ripples of all sizes, and you can measure the intensity of ripples of different sizes. And you can discuss what we call the spectrum—we use that word exactly the way it's used to describe sound waves. When we talk about the spectrum of a sound wave, we're talking about how the intensity varies with the different wavelengths that make up that sound wave. We do exactly the same thing in the early universe, and talk about how the intensity of these ripples in the mass density of the early universe varied with the wavelengths of the different ripples we're looking at. Today we can see those ripples in the cosmic background radiation.

The fact that we can see them at all is an absolutely fantastic success of modern technology. When we were first making these predictions, back in 1982, at that time astronomers had just barely been able to see the effect of the Earth's motion through the cosmic background radiation, which is an effect of about one part in a thousand. The ripples I'm talking about are only one part in 100,000—just 1 percent of the intensity of the most subtle effect it had been possible to observe at the time we were first doing these calculations.

I never believed we would ever actually see these ripples. It just seemed too far-fetched that astronomers would get to be a hundred times better at measuring these things than they were at the time. But to my astonishment and delight, in 1992 these ripples were first detected by a satellite called COBE, the Cosmic Background Explorer, and now we have far better measurements than COBE, which had an angular resolution of about 7 degrees. This meant that you could see only the longest wavelength ripples. Now we have measurements that go down to a fraction of a degree, and we're getting very precise measurements now of how the intensity varies with wavelength, with marvelous success.

About a year and a half ago, there was a spectacular set of announcements from experiments called BOOMERANG and MAXIMA, both balloon-based experiments, which gave very strong evidence that the universe is geometrically flat, which is just what inflation predicts. By flat I don't mean two-dimensional; I just mean that the three-dimensional space of the universe is not curved, as it could have been according to general relativity. You can actually see the curvature of space in the way that the pattern of ripples has been affected by the evolution of the universe. A year and a half ago, however, there was an important discrepancy that people worried about, and no one was sure how big a deal to make out of it. The spectrum they were measuring was a graph that had, in principle, several peaks. These peaks had to do with successive oscillations of the density waves in the early universe and a phenomenon called resonance that makes some wavelengths more intense than others. The measurements showed the first peak beautifully, exactly where we expected it to be, with just the shape that was expected. But we couldn't actually see the second peak.

In order to fit the data with the theories, people had to assume there were about ten times as many protons in the universe as we actually thought, because the extra protons would lead to a friction effect that could make the second peak disappear. Of course, every experiment has some uncertainty—if an experiment is performed many times, the results will not be exactly the same each time. So we could imagine that the second peak was not seen purely because of bad luck. However, the probability that the peak could be so invisible, if the universe contained the density of protons that's indicated by other measurements, was down to about the 1-percent level. So it was a very serious-looking discrepancy between what was observed and what was expected.

All this changed dramatically for the better about three or four months ago, with the next set of announcements with more precise measurements. Now the second peak is not only visible but it has exactly the height that was expected, and everything about the data now fits beautifully with the theoretical predictions. Too good, really. I'm sure it will get worse before it continues to get

Alan Guth

better, given the difficulties in making these kinds of measurements. But we have a beautiful picture now which seems to be confirming the inflationary theory of the early universe.

Our current picture of the universe has a new twist, however, which was discovered two or three years ago. To make things fit, to match the observations, which are now getting very clear, we have to assume that there's a new component of energy in the universe which we didn't know existed before. This new component is usually referred to as dark energy. As the name clearly suggests, we still don't know exactly what this new component is. It's a component of energy which in fact is very much like the repulsive gravity matter I talked about earlier—the material that drives the inflation in the early universe. It appears that, in fact, today the universe is filled with a similar kind of matter. The antigravity effect is much weaker than the effect I was talking about in the early universe, but the universe today appears very definitely to be starting to accelerate again, under the influence of this so-called dark energy.

Although I'm trying to advertise that we've understood a lot, and we have, there are still many uncertainties. In particular, we still don't know what most of the universe is made out of. There's the dark energy, which seems to comprise in fact about 60 percent of the total mass/energy of the universe. We don't know what it is. It could in fact be the energy of the vacuum itself, but we don't know that for a fact. In addition, there's what we call dark matter, which is another 30 percent, or maybe almost 40 percent, of the total matter in the universe. We don't know what that is, either. The difference between the two is that the dark energy causes repulsive gravity and is smoothly distributed; the dark matter behaves like ordinary matter in terms of its gravitational properties—it's attractive and it clusters, but we don't know what it's made of. The stuff we do know about—protons, neutrons, ordinary atoms and molecules—appear to comprise only about 5 percent of the mass of the universe.

The moral of the story is we have a great deal to learn. At the same time, the theories that we have developed so far seem to be working almost shockingly well.

2
The Cyclic Universe

Paul Steinhardt

Theoretical physicist; Albert Einstein Professor of Science, Princeton University; coauthor (with Neil Turok), *Endless Universe: Beyond the Big Bang*

If you were to ask most cosmologists to give a summary of where we stand right now in the field, they would tell you that we live in a very special period in human history where, thanks to a whole host of advances in technology, we can suddenly view the very distant and very early universe in ways we haven't been able to do ever before. For example, we can get a snapshot of what the universe looked like in its infancy, when the first atoms were forming. We can get a snapshot of what the universe looked like in its adolescence, when the first stars and galaxies were forming. And we are now getting a full detail, three-dimensional image of what the local universe looks like today. When you put together this different information, which we're getting for the first time in human history, you obtain a very tight series of constraints on any model of cosmic evolution.

If you go back to the different theories of cosmic evolution in the early 1990s, the data we've gathered in the last decade has eliminated all of them save one, a model that you might think of today as the consensus model. This model involves a combination of the Big Bang model as developed in the 1920s, '30s, and '40s; the inflationary theory, which Alan Guth proposed in the 1980s; and a recent amendment that I will discuss shortly. This consensus theory matches the observations we have of the universe today in exquisite detail. For this reason, many cosmologists conclude that we have finally determined the basic cosmic history of the universe.

But I have a rather different point of view, a view that has been stimulated by two events. The first is the recent amendment to which I referred earlier. I want to argue that the recent amendment is not simply an amendment but a real shock to our whole notion of time and cosmic history. And secondly, in the last year I've been involved in the development of an alternative theory that turns the cosmic history topsy-turvy: All the events that created the important features of our universe occur in a different order, by different physics, at different times, over different time scales. And yet this model seems capable of reproducing all the successful predictions of the consensus picture with the same exquisite detail.

The key difference between this picture and the consensus picture comes down to the nature of time. The standard model, or consensus model, assumes that time has a beginning that we normally refer to as the Big Bang. According to that model, for reasons we don't quite understand, the universe sprang from nothingness into somethingness, full of matter and energy, and has been expanding and cooling for the past 15 billion years. In the alternative model, the universe is endless. Time is endless, in the sense that it goes on forever in the past and forever in the future, and in some sense space is endless. Indeed, our three spatial dimensions remain infinite throughout the evolution of the universe.

More specifically, this model proposes a universe in which the evolution of the universe is cyclic. That is to say, the universe goes through periods of evolution from hot to cold, from dense to under-dense, from hot radiation to the structure we see today, and eventually to an empty universe. Then, a sequence of events occurs that cause the cycle to begin again. The empty universe is reinjected with energy, creating a new period of expansion and cooling. This process repeats periodically forever. What we're witnessing now is simply the latest cycle.

The notion of a cyclic universe is not new. People have considered this idea as far back as recorded history. The ancient Hindus, for example, had a very elaborate and detailed cosmology based on a cyclic universe. They predicted the duration of each cycle to be 8.64 billion years—a prediction with three-digit accuracy. This is very impressive, especially since they had no quantum mechanics and no string theory! It disagrees with the number I'm going suggest, which is trillions of years rather than billions.

The cyclic notion has also been a recurrent theme in Western thought. Edgar Allan Poe and Friedrich Nietzsche, for example, each had cyclic models of the universe, and in the early days of relativistic cosmology Albert Einstein, Alexander Friedmann, Georges Lemaître, and Richard Tolman were interested in the cyclic idea. I think it's clear why so many have found the cyclic idea to be appealing: If you have a universe with a beginning, you have the challenge of explaining why it began and the conditions under which it began. If you have a universe that's cyclic, it's eternal, so you don't have to explain the beginning.

During the attempts to try to bring cyclic ideas into modern cosmology, it was discovered in the 1920s and '30s that there are various technical problems. The idea at that time was a cycle in which our three-dimensional universe goes through periods of expansion beginning from the Big Bang and then reversal to contraction and a Big Crunch. The universe bounces, and expansion begins again. One problem is that every time the universe contracts to a crunch, the density and temperature of the universe rises to an infinite value, and it is not clear if the usual laws of physics can be applied.

Second, every cycle of expansion and contraction creates entropy through natural thermodynamic processes, which adds to the entropy from earlier cycles. So at the beginning of a new cycle, there is higher entropy density than the cycle before. It turns out that the duration of a cycle is sensitive to the entropy density. If the entropy increases, the duration of the cycle increases as well. So, going forward in time, each cycle becomes longer than the one before. The problem is that, extrapolating back in time, the cycles become shorter until, after a finite time, they shrink to zero duration. The problem of avoiding a beginning has not been solved; it has simply been pushed back a finite number of cycles. If we're going to reintroduce the idea of a truly cyclic universe, these two problems must be overcome. The cyclic model I will describe uses new ideas to do just that.

To appreciate why an alternative model is worth pursuing, it's important to get a more detailed impression of what the consensus picture is like. Certainly some aspects are appealing. But what I want to argue is that, overall, the consensus model is not so simple. In particular, recent observations have forced us to amend the consensus model and make it more complicated. So, let me begin with an overview of the consensus model.

The consensus theory begins with the Big Bang: The universe has a beginning. It's a standard assumption that people have made over the last fifty years, but it's not something we can prove at present from any fundamental laws of physics. Furthermore, you have to assume that the universe began with an energy density less than the critical value. Otherwise, the universe would stop expanding and recollapse before the next stage of evolution, the inflationary epoch. In addition, to reach this inflationary stage, there must be some sort of energy to drive the inflation. Typically this is assumed to be due to an inflation field. You have to assume that in those patches of the universe that began at less than the critical density, a significant fraction of the energy is stored in inflation energy so that it can eventually overtake the universe and start the period of accelerated expansion. All of these are reasonable assumptions, but assumptions nevertheless. It's important to take into account these assumptions and ingredients, because they're helpful in comparing the consensus model to the challenger.

Assuming these conditions are met, the inflation energy overtakes the matter and radiation after a few instants. The inflationary epoch commences, and the expansion of the universe accelerates at a furious pace. The inflation does a number of miraculous things: It makes the universe homogeneous, it makes the universe flat, and it leaves behind certain inhomogeneities, which are supposed to be the seeds for the formation of galaxies. Now the universe is prepared to enter the next stage of evolution with the right conditions. According to the inflationary model, the inflation energy decays into a hot gas of matter and radiation. After a second or so, there form the first light nuclei. After a few tens of thousands of years, the slowly moving matter dominates the universe. It's during these stages that the first atoms form, the universe becomes transparent, and the structure in the universe begins to form—the first stars and galaxies. Up to this point, the story is relatively simple.

But there is the recent discovery that we've entered a new stage in the evolution of the universe. After the stars and galaxies have formed, something strange has happened to cause the expansion of the universe to speed up again. During the 15 billion years when matter and radiation dominated the universe and structure was forming, the expansion of the universe was slowing down, because the matter and radiation within it is gravitationally self-attractive and resists the expansion of the universe. Until very recently, it had been presumed that matter would continue to be the dominant form of energy in the universe and this deceleration would continue forever.

But we've discovered instead, due to recent observations, that the expansion of the universe is speeding up. This means that most of the energy of the universe is neither matter nor radiation. Rather, another form of energy has overtaken the matter and radiation. For lack of a better term, this new energy form is called dark energy. Dark energy, unlike the matter and radiation we're familiar with, is gravitationally self-repulsive. That's why it causes the expansion to speed up rather than slow down. In Newton's theory of gravity, all mass is gravitationally attractive, but Einstein's theory allows the possibility of forms of energy that are gravitationally self-repulsive.

I don't think either the physics or cosmology communities, or even the general public, have fully absorbed the full implications of this discovery. This is a revolution in the grand historic sense—in the Copernican sense. In fact, if you think about Copernicus— from whom we derive the word "revolution"—his importance was that he changed our notion of space and of our position in the universe. By showing that the Earth revolves around the sun, he triggered a chain of ideas that led us to the notion that we live in no particular place in the universe; there's nothing special about where we are. Now we've discovered something very strange about the nature of time: that we may live in no special place, but we do live at a special time, a time of recent transition from deceleration to acceleration; from one in which matter and radiation dominate the universe to one in which they are rapidly becoming insignificant components; from one in which structure

Paul Steinhardt

is forming in ever larger scales to one in which now, because of this accelerated expansion, structure formation stops. We are in the midst of the transition between these two stages of evolution. And just as Copernicus' proposal that the Earth is no longer the center of the universe led to a chain of ideas that changed our whole outlook on the structure of the solar system and eventually to the structure of the universe, it shouldn't be too surprising that perhaps this new discovery of cosmic acceleration could lead to a whole change in our view of cosmic history. That's a big part of the motivation for thinking about our alternative proposal.

With these thoughts about the consensus model in mind, let me turn to the cyclic proposal. Since it's cyclic, I'm allowed to begin the discussion of the cycle at any point I choose. To make the discussion parallel, I'll begin at a point analogous to the Big Bang; I'll call it the Bang. This is a point in the cycle where the universe reaches its highest temperature and density. In this scenario, though, unlike the Big Bang model, the temperature and density don't diverge. There is a maximal, finite temperature. It's a very high temperature, around 10^{20} degrees Kelvin—hot enough to evaporate atoms and nuclei into their fundamental constituents—but it's not infinite. In fact, it's well below the so-called Planck energy scale, where quantum gravity effects dominate. The theory begins with a bang and then proceeds directly to a phase dominated by radiation. In this scenario you do not have the inflation one has in the standard scenario. You still have to explain why the universe is flat, you still have to explain why the universe is homogeneous, and you still have to explain where the fluctuations came from that led to the formation of galaxies, but that's not going to be explained by an early stage of inflation. It's going to be explained by yet a different stage in the cyclic universe, which I'll get to.

In this new model, you go directly to a radiation-dominated universe and form the usual nuclear abundances; then go directly to a matter-dominated universe in which the atoms and galaxies and larger-scale structure form; and then proceed to a phase of the universe dominated by dark energy. In the standard case, the dark energy comes as a surprise, since it's something you have to add into the theory to make it consistent with what we observe. In the cyclic model, the dark energy moves to center stage as the key ingredient that is going to drive the universe, and in fact drives the universe, into the cyclic evolution. The first thing the dark energy does when it dominates the universe is what we observe today: It causes the expansion of the universe to begin to accelerate. Why is that important? Although this acceleration rate is 100 orders of magnitude smaller than the acceleration that one gets in inflation, if you give the universe enough time it actually accomplishes the same feat that inflation does. Over time, it thins out the distribution of matter and radiation in the universe, making the universe more and more homogeneous and isotropic—in fact, making it perfectly so—driving it into what is essentially a vacuum state.

Seth Lloyd said there were 10^{80} or 10^{90} bits inside the horizon, but if you were to look around the universe in a trillion years, you would find on average no bits inside your horizon, or less than one bit inside your horizon. In fact, when you count these bits, it's important to realize that now that the universe is accelerating, our computer is actually losing bits from inside our horizon. This is something that we observe.

At the same time that the universe is made homogeneous and isotropic, it is also being made flat. If the universe had any warp or curvature to it, or if you think about the universe stretching over this long period of time, although it's a slow process it

Paul Steinhardt

makes the space extremely flat. If it continued forever, of course, that would be the end of the story. But in this scenario, just like inflation, the dark energy survives only for a finite period and triggers a series of events that eventually lead to a transformation of energy from gravity into new energy and radiation that will then start a new period of expansion of the universe. From a local observer's point of view, it looks like the universe goes through exact cycles; that is to say, it looks like the universe empties out each round and a new matter and radiation is created, leading to a new period of expansion. In this sense it's a cyclic universe. If you were a global observer and could see the entire universe, you'd discover that our three dimensions are forever infinite in this story. What's happened is that at each stage when we create matter and radiation, it gets thinned out. It's out there somewhere, but it's getting thinned out. Locally, it looks like the universe is cyclic, but globally the universe has a steady evolution, a well-defined era in which, over time and throughout our three dimensions, entropy increases from cycle to cycle.

Exactly how this works in detail can be described in various ways. I will choose to present a very nice geometrical picture that's motivated by superstring theory. We use only a few basic elements from superstring theory, so you don't really have to know anything about superstring theory to understand what I'm going to talk about, except to understand that some of the strange things I'm going to introduce I am not introducing for the first time. They're already sitting there in superstring theory waiting to be put to good purpose.

One of the ideas in superstring theory is that there are extra dimensions; it's an essential element to that theory, which is necessary to make it mathematically consistent. In one particular formulation of that theory, the universe has a total of eleven dimensions. Six of them are curled up into a little ball so tiny that, for my purposes, I'm just going to pretend they're not there. However, there are three spatial dimensions, one time dimension, and one additional dimension that I do want to consider. In this picture, our three dimensions with which we're familiar and through which we move lie along a hypersurface, or membrane. This membrane is a boundary of the extra dimension. There is another boundary, or membrane, on the other side. In between, there's an extra dimension that, if you like, only exists over a certain interval. It's like we are one end of a sandwich, in between which there is a so-called bulk volume of space. These surfaces are referred to as orbifolds or branes—the latter referring to the word "membrane." The branes have physical properties. They have energy and momentum, and when you excite them you can produce things like quarks and electrons. We are composed of the quarks and electrons on one of these branes. And, since quarks and leptons can only move along branes, we are restricted to moving along and seeing only the three dimensions of our brane. We cannot see directly the bulk or any matter on the other brane.

In the cyclic universe, at regular intervals of trillions of years, these two branes smash together. This creates all kinds of excitations—particles and radiation. The collision thereby heats up the branes, and then they bounce apart again. The branes are attracted to each other through a force that acts just like a spring, causing the branes to come together at regular intervals. To describe it more completely, what's happening is that the universe goes through two kinds of stages of motion. When the universe has matter and radiation in it, or when the branes are far enough apart, the main motion is the branes stretching, or, equivalently, our three dimensions expanding. During this period, the branes more or less remain a fixed distance apart. That's what's been happening, for example, in the last 15 billion years. During these stages, our three dimensions are stretching just as they normally would. At a microscopic distance away, there is another brane sitting and expanding, but since we can't touch, feel, or see across the bulk, we can't sense it directly. If there is a clump of matter over there, we can feel the gravitational effect, but we can't see any light or anything else it emits, because anything it emits is going to move along that brane. We only see things that move along our own brane.

Next, the energy associated with the force between these branes takes over the universe. From our vantage point on one of the branes, this acts just like the dark energy we observe today. It causes the branes to accelerate in their stretching, to the point where all the matter and radiation produced since the last collision is spread out and the branes become essentially smooth, flat, empty surfaces. If you like, you can think of them as being wrinkled and full of matter up to this point, and then stretching by a fantastic amount over the next trillion years. The stretching causes the mass and energy on the brane to thin out and the wrinkles to be smoothed out. After trillions of years, the branes are, for all intents and purposes, smooth, flat, parallel, and empty.

Then the force between these two branes slowly brings the branes together. As it brings them together, the force grows stronger and the branes speed toward one another. When they collide, there's a walloping impact—enough to create a high density of matter and radiation with a very high, albeit finite, temperature. The two branes go flying apart, more or less back to where they are, and then the new matter and radiation, through the action of gravity, causes the branes to begin a new period of stretching.

In this picture, it's clear that the universe is going through periods of expansion and a funny kind of contraction. Where the two branes come together, it's not a contraction of our dimensions but a contraction of the extra dimension. Before the contraction, all matter and radiation has been spread out, but, unlike the old cyclic models of the 1920s and '30s, it doesn't come back together again during the contraction, because our three dimensions—that is, the branes—remain stretched out. Only the extra dimension contracts. This process repeats itself cycle after cycle.

If you compare the cyclic model to the consensus picture, two of the functions of inflation—namely, flattening and homogenizing the universe—are accomplished by the period of accelerated expansion that we've now just begun. Of course, I really mean the analogous expansion that occurred one cycle ago, before the most recent Bang. The third function of inflation—producing fluctuations in the density—occurs as these two branes come together. As they approach, quantum fluctuations cause the branes to begin to wrinkle. And because they're wrinkled, they don't collide everywhere at the same time. Rather, some regions collide a bit earlier than others. This means that some regions reheat to a finite temperature and begin to cool a little bit before other regions. When the branes come apart again, the temperature of the universe is not perfectly homogeneous but has spatial variations left over from the quantum wrinkles.

Remarkably, although the physical processes are completely different and the time scale is completely different—this is taking billions of years, instead of 10^{-30} seconds—it turns out that the spectrum of fluctuations you get in the distribution of energy and temperature is essentially the same as what you get in inflation. Hence, the cyclic model is also in exquisite agreement with all of the measurements of the temperature and mass distribution of the universe that we have today.

Paul Steinhardt

Because the physics in these two models is quite different, there is an important distinction in what we would observe if one or the other were actually true—although this effect has not been detected yet. In inflation when you create fluctuations, you don't just create fluctuations in energy and temperature but you also create fluctuations in spacetime itself, so-called gravitational waves. That's a feature we hope to look for in experiments in the coming decades as a verification of the consensus model. In our model, you don't get those gravitational waves. The essential difference is that inflationary fluctuations are created in a hyper-rapid, violent process that is strong enough to create gravitational waves, whereas cyclic fluctuations are created in an ultraslow, gentle process that is too weak to produce gravitational waves. That's an example where the two models give an observational prediction that is dramatically different. It's just difficult to observe at the present time.

What's fascinating at the moment is that we have two paradigms now available to us. On the one hand, they are poles apart in terms of what they tell us about the nature of time, about our cosmic history, about the order in which events occur, and about the time scale on which they occur. On the other hand, they are remarkably similar in terms of what they predict about the universe today. Ultimately what will decide between the two is a combination of observations—for example, the search for cosmic gravitational waves—and theory, because a key aspect to this scenario entails assumptions about what happens at the collision between branes that might be checked or refuted in superstring theory. In the meantime, for the next few years, we can all have great fun speculating about the implications of each of these ideas and how we can best distinguish between them.

3
The Inflationary Universe

Alan Guth

Paul Steinhardt did a very good job of presenting the case for the cyclic universe. I'm going to describe the conventional consensus model upon which he was trying to say that the cyclic model is an improvement. I agree with what Paul said at the end of his talk about comparing these two models; it's yet to be seen which one works. But there are two grounds for comparing them. One is that in both cases the theory needs to be better developed. This is more true for the cyclic model, where one has the issue of what happens when branes collide. The cyclic theory could die when that problem finally gets solved definitively. Secondly, there is, of course, the observational comparison of the gravitational-wave predictions of the two models.

A brane is short for "membrane," a term that comes out of string theories. String theories began purely as theories of strings, but when people began to study their dynamics more carefully, they discovered that for consistency it was not possible to have a theory which discussed only strings. Whereas a string is a one-dimensional object, the theory also had to include the possibility of membranes of various dimensions to make it consistent, which led to the notion of branes in general. The theory that Paul described in particular involves a four-dimensional space plus one time dimension, which he called the bulk. That four-dimensional space was sandwiched between two branes.

That's not what I'm going to talk about. I want to talk about the conventional inflationary picture, and in particular the great boost that this picture has attained over the past few years by the somewhat shocking revelation of a new form of energy that exists in the universe. This energy, for lack of a better name, is typically called dark energy.

But let me start the story further back. Inflationary theory itself is a twist on the conventional Big Bang theory. The short-coming that inflation is intended to overcome is the basic fact that although the Big Bang theory is called the Big Bang, it is in fact not really a theory of a bang at all; it never was. The conventional Big Bang theory, without inflation, was really only a theory of the aftermath of the Bang. It started with all of the matter in the universe already in place, already undergoing rapid expansion, already incredibly hot. There was no explanation of how it got that way. Inflation is an attempt to answer that question, to say what "banged," and what drove the universe into this period of enormous expansion. Inflation does that very wonderfully. It explains not only what caused the universe to expand but also the origin of essentially all the matter in the universe at the same time. I qualify that with the word "essentially" because, in a typical version of the theory, inflation needs about a gram's worth of matter to start. So inflation is not quite a theory of the ultimate beginning, but it is a theory of evolution that explains essentially everything we see around us, starting from almost nothing.

The basic idea behind inflation is that a repulsive form of gravity caused the universe to expand. General relativity, from its inception, predicted the possibility of repulsive gravity; in the context of general relativity, you basically need a material with a negative pressure to create repulsive gravity. According to general relativity, it's not just matter densities or energy densities that create gravitational fields, it's also pressures. A positive pressure creates a normal attractive gravitational field, of the kind we're accustomed to, but a negative pressure would create a repulsive kind of gravity. It also turns out that according to modern particle theories, materials with a negative pressure are easy to construct out of fields that exist according to these theories. By putting together these two ideas—the fact that particle physics gives us states with negative pressures, and that general relativity tells us that those states cause a gravitational repulsion—we reach the origin of the inflationary theory.

By answering the question of what drove the universe into expansion, the inflationary theory can also answer some questions about that expansion that would otherwise be mysterious. There are two very important properties of our observed universe that were never really explained by the Big Bang theory; they were just part of one's assumptions about the initial conditions. One of them is the uniformity of the universe—the fact that it looks the same everywhere, no matter which way you look, as long as you average over large enough volumes. It's both isotropic, meaning the same in all directions, and homogeneous, meaning the same in all places. The conventional Big Bang theory never really had an explanation for that; it just had to be assumed from the start. The problem is that although we know that any set of objects will approach a uniform temperature if they're allowed to sit for a long time, the early universe evolved so quickly that there wasn't enough time for this to happen. To explain, for example, how the universe could have smoothed itself out to achieve the uniformity of temperature we observe today in the cosmic background radiation, one finds that in the context of the standard Big Bang theory it would be necessary for energy and information to be transmitted across the universe at about a hundred times the speed of light.

In the inflationary theory, this problem goes away completely, because, in contrast to the conventional theory, it postulates a period of accelerated expansion while this repulsive gravity is taking place. That means that if we follow our universe backward in time toward the beginning using inflationary theory, we see that it started from something much smaller than you ever could have imagined in the context of conventional cosmology without inflation. While the region that would evolve to become our universe was incredibly small, there was plenty of time for it to reach a uniform temperature, just like a cup of coffee sitting on the table cools down to room temperature. Once this uniformity is established on this tiny scale by normal thermal-equilibrium processes—and I'm talking now about something that's about a billion times smaller than the size of a single proton—inflation can take over and cause this tiny region to expand rapidly and become large enough to encompass the entire visible universe. The inflationary theory not only allows the possibility for the universe to be uniform but also tells us why it's uniform: It's uniform because it came from something that had time to become uniform and was then stretched by the process of inflation.

The second peculiar feature of our universe that inflation does a wonderful job of explaining, and for which there never was a prior explanation, is the flatness of the universe—the fact that the geometry of the universe is so close to Euclidean. In the context of relativity, Euclidean geometry is not the norm, it's an oddity. With general relativity, curved space is the generic case. In the case of the universe as a whole, once we assume that the universe is homogeneous and isotropic, then this issue of flatness becomes directly related to the relationship between the mass density and the expansion rate of the universe. A large mass density would cause space to curve into a closed universe in the shape of a ball;

if the mass density dominated, the universe would be a closed space with a finite volume and no edge. If a spaceship traveled in what it thought was a straight line for a long enough distance, it would end up back where it started from. In the alternative case, if the expansion dominated, the universe would be geometrically open. Geometrically open spaces have the opposite geometric properties from closed spaces. They're infinite. In a closed space, two lines which are parallel will start to converge; in an open space, two lines which are parallel will start to diverge. In either case, what you see is very different from Euclidean geometry. However, if the mass density is right at the borderline of these two cases, then the geometry is Euclidean, just like we all learned about in high school.

In terms of the evolution of the universe, the fact that the universe is at least approximately flat today requires that the early universe was extraordinarily flat. The universe tends to evolve away from flatness, so even given what we knew ten or twenty years ago—we know much better, now, that the universe is extraordinarily close to flat—we could have extrapolated backward and discovered that, for example, at one second after the Big Bang the mass density of the universe must have been equal, to an accuracy of fifteen decimal places, to the critical density where it counterbalanced the expansion rate to produce a flat universe. The conventional Big Bang theory gave us no reason to believe that there was any mechanism to require that, but it has to have been that way, to explain why the universe looks the way it does today. The conventional Big Bang theory without inflation really only worked if you fed into it initial conditions which were highly finely tuned to make it just right to produce a universe like the one we see. Inflationary theory gets around this flatness problem, because inflation changes the way the geometry of the universe evolves with time. Even though the universe always evolves away from flatness at all other periods in the history of the universe, during the inflationary period the universe is actually driven towards flatness incredibly quickly. If you had approximately 10^{-34} seconds or so of inflation at the beginning of the universe, that's all you need. Inflation would then have driven the universe to be flat closely enough to explain what we see today.

There are two primary predictions that come out of inflationary models, which appear to be testable today. They have to do (1) with the mass density of the universe, and (2) with the properties of the density nonuniformities. I'd like to say a few words about each of them, one at a time. Let me begin with the question of flatness.

The mechanism that inflation provides that drives the universe toward flatness will in almost all cases overshoot, not giving us a universe that is just nearly flat today but a universe that's almost *exactly* flat today. This can be avoided, and people have at times tried to design versions of inflation that avoided it, but those versions of inflation never looked very plausible. You have to arrange for inflation to end at just the right point, where it's almost made the universe flat but not quite. It requires a lot of delicate fine-tuning, but in the days when it looked like the universe was open, some people tried to design such models. But they always looked contrived and never really caught on.

The generic inflationary model drives the universe to be completely flat, which means that one of the predictions is that today the mass density of the universe should be at the critical value that makes the universe geometrically flat. Until three or four years ago, no astronomers believed that. They told us that if you looked at just the visible matter, you would see only about 1 percent of what you needed to make the universe flat. But they also said they could offer more than that: There's also dark matter. Dark matter is matter that's inferred to exist because of the gravitational effect it has on visible matter. It's seen, for example, in the rotation curves of galaxies. When astronomers first measured how fast galaxies rotate, they found they were spinning so fast that if the only matter present was what you saw, galaxies would just fly apart.

To understand the stability of galaxies, it was necessary to assume that there was a large amount of dark matter in the galaxy—about five or ten times the amount of visible matter— which was needed just to hold the galaxy together. This problem repeats itself when one talks about the motion of galaxies within clusters of galaxies. The motion of galaxies in clusters is much more random and chaotic than the spiral galaxy, but the same issues arise. You can ask how much mass is needed to hold those clusters of galaxies together, and the answer is that you still need significantly more matter than what you assumed was in the galaxies. Adding all of that together, astronomers came up only with about a third of the critical density. They were pretty well able to guarantee that there wasn't any more than that out there; that was all they could detect. That was bad for the inflationary model, but many of us still had faith that inflation had to be right and that sooner or later the astronomers would come up with something.

And they did, although what they came up with was something very different from the kind of matter we were talking about previously. Starting in 1998, astronomers have been gathering evidence for the remarkable fact that the universe today appears to be accelerating, not slowing down. As I said at the beginning of this talk, the theory of general relativity allows for that. What's needed is a material with a negative pressure. We are now therefore convinced that our universe must be permeated with a material with negative pressure which is causing the acceleration we're now seeing. We don't know what this material is, but we're referring to it as dark energy. Even without knowing what it is, general relativity by itself allows us to calculate how much mass has to be out there to cause the observed acceleration, and it turns out to be almost exactly equal to two-thirds of the critical density. This is exactly what was missing from the previous calculations! So, if we assume that this dark energy is real, we now have complete agreement between what the astronomers are telling us about the mass density of the universe and what inflation predicts.

The other important prediction that comes out of inflation is becoming even more persuasive than the issue of flatness: namely, the issue of density perturbations. Inflation has what in some ways is a wonderful characteristic: that by stretching everything out—and Paul's model takes advantage of the same effect—you can smooth out any nonuniformities that were present prior to this expansion. Inflation does not depend sensitively on what you assume existed before inflation; everything there just gets washed away by the enormous expansion. For a while, in the early days of developing the inflationary model, we were all very worried that this would lead to a universe that would be absolutely, completely smooth.

After a while, several physicists began to explore the idea that quantum fluctuations could save us. The universe is fundamentally a quantum mechanical system, so perhaps quantum theory was necessary not just to understand atoms but also to understand galaxies. It's a rather remarkable idea that an aspect of fundamental physics like quantum theory could have such a broad sweep. The point is that a classical version of inflationary theory would predict a completely uniform density of matter at the end of inflation. According to quantum mechanics, however, everything is probabilistic. There are quantum fluctuations everywhere, which means that in some places the mass density would be slightly higher than average and in other places it would be slightly lower than average. That's exactly the sort of thing you want, to explain the structure of the universe. You can even go ahead and calculate the spectrum of these nonuniformities, which is something that Paul and I both worked on in the early days and had great fun with. The answer that we both came up with was that, in fact, quantum mechanics produces just the right spectrum of nonuniformities.

We really can't predict the overall amplitude—that is, the intensity—of these ripples unless we know more about the fundamental theory. At the present time, we have to take the overall factor that multiplies the predicted intensity of these ripples from observation. But we can predict the spectrum—that is, the complicated pattern of ripples can be viewed as ripples of many different wavelengths lying on top of each other, and we can calculate how the intensity of the ripples varies with their wavelengths. We knew how to do this back in 1982, but recently it has actually become possible for astronomers to see these nonuniformities imprinted on the cosmic background radiation. These were first observed back in 1992 by the COBE satellite, but back then they could only see very broad features, since the angular resolution of the satellite was only about 7 degrees. Now they've gotten down to angular resolutions of about 1/10 of a degree. These observations of the cosmic background radiation can be used to produce plots of the spectrum of nonuniformities, which are becoming more and more detailed.

The most recent data set was made by an experiment called the Cosmic Background Imager, which released a new set of data in May that is rather spectacular. This graph of the spectrum is rather complicated, because these fluctuations are produced during the inflationary era but then oscillate as the early universe evolves. Thus, what you see is a picture that includes the original spectrum plus all of the oscillations that depend on various properties of the universe. A remarkable thing is that these curves now show five separate peaks, and all five of the peaks show good agreement between theory and observation. You can see that the peaks are in about the right place and have about the right heights, without any ambiguity, and the leading peak is rather well-mapped-out. It's a rather remarkable fit between actual measurements made by astronomers and a theory based on wild ideas about quantum fluctuations at 10^{-35} seconds. The data are so far in beautiful agreement with the theory.

At the present time, this inflationary theory, which a few years ago was in significant conflict with observation, now works perfectly with our measurements of the mass density and the fluctuations. The evidence for a theory that's either the one I'm talking about or something very close to it is very, very strong.

I'd just like to close by saying that although I've been using "theory" in the singular to talk about inflation, I shouldn't, really. It's important to remember that inflation is really a class of theories. If inflation is right, it's by no means the end of our study of the origin of the universe, but still it's really closer to the beginning. There are many different versions of inflation, and in fact the cyclic model that Paul described could be considered one version. It's a rather novel version, since it puts the inflation at a completely different era of the history of the universe, but inflation is still doing many of the same things. There are many versions of inflation that are much closer to the kinds of theories we were developing in the '80s and '90s, so saying that inflation is right is by no means the end of the story. There's still a lot of flexibility here, and a lot to be learned. And what needs to be learned will involve both the study of cosmology and the study of the underlying particle physics, which is essential to these models.

Alan Guth

4
A Balloon Producing Balloons Producing Balloons

Andrei Linde

Theoretical physicist, Stanford University; father of eternal chaotic inflation; inaugural winner, Milner Foundation Fundamental Physics Prize

I should probably start by explaining what happened during the last thirty years in cosmology. This story will begin with old news: the creation of inflationary theory. Then we will talk about the relatively recent developments, when inflation became a part of the theory of an inflationary multiverse and the string theory landscape. Then—what we expect in the future.

Let me start by saying that many, many years ago—and I mean like almost a century ago—Einstein came up with something called the cosmological principle, which says that our universe must be homogeneous and uniform. And for many years people used this principle. In fact, it was formulated even much earlier, by Newton. The universe is still represented this way in current books on astrophysics, where you can find different versions of the cosmological principle.

For a while, this was the only way of answering the question of why the universe is everywhere the same—in fact, why it is the universe. So we did not think about the multiverse, we just wanted to explain why the world is so homogeneous around us, why it is so big, why parallel lines do not intersect. Which is, in fact, part of the same question: If the universe was tiny, like a small globe, and you drew parallel lines perpendicular to the equator of the globe, they would intersect at the south and the north poles. Why has nobody ever seen parallel lines intersecting?

These kinds of questions, for many years, could seem a bit silly. For example, one may wonder what happened before the universe even emerged. The textbook of general relativity that we used in Russia said that it was meaningless to ask this question, because the solutions of the Einstein equations cannot be continued through the singularity, so why bother? And yet people bothered. They are still trying to answer these kinds of questions. But for many people such questions looked metaphysical, not to be taken seriously.

When inflationary theory was invented, people started taking these questions seriously. Alan Guth asked these questions and proposed the theory of cosmic inflation, a framework in which a consistent answer to these questions could be found. The problem was, as Guth immediately recognized, that his own answer to these questions was incomplete. And then, after more than a year of work, I proposed a new version of inflationary theory, which helped to find a way to answer many of these questions. At first it sounded like science fiction, but once we found possible answers to the questions which previously were considered metaphysical, we couldn't just forget about it. This was the first reason for us to believe the idea of cosmic inflation. So let me explain this idea.

Standard Big Bang theory says that everything begins with a big bang, a huge explosion. Terrorists started the universe. But when you calculate how much high-tech explosives these guys would have to have at their disposal to start the universe formation, they would need 10^{80} tons of high-tech explosives, compressed to a ball smaller than 1 centimeter, and ignite all of its parts exactly at the same time with a precision better than 1 in 10,000.

Andrei Linde

Another problem was that in the standard Big Bang scenario, the universe could only expand slower as time went on. But then why did the universe start to expand? Who gave it the first push? It looks totally incredible, like a miracle. However, people sometimes believe that the greater the miracle, the better: Obviously, God could create 10^{80} tons of explosive from nothing, then ignite it and make it grow, all for our benefit. Can we make an attempt to come up with an alternative explanation?

According to inflationary theory, one may avoid many of these problems if the universe began in some special state, almost a vacuum-like state. The simplest version of such a state involves something called a scalar field. Remember electric and magnetic fields? Well, a scalar field is even simpler: It doesn't point to any direction. If it's uniform and doesn't change in time, it is invisible like a vacuum, but it may have lots of energy packed in it. When the universe expands, the scalar field remains almost constant, and its energy density remains almost constant.

This is the key point, so let us talk about it. Think about the universe as a big box containing many atoms. When the universe expands two times, its volume grows eight times, and therefore the density of atoms decreases eight times. However, when the universe is filled with a constant scalar field, its energy density remains constant while the universe expands. Therefore when the size of the universe grows two times, the total energy of matter in the universe grows eight times. If the universe continues to grow, its total energy, and its total mass, rapidly becomes enormously large, so one could easily get all of these 10^{80} tons of matter starting from almost nothing.

That was the basic idea of inflation. At the first glance, it could seem totally wrong, because of energy conservation. One cannot get energy from nothing. We always have the same energy with which we started. Once, I was invited to give an opening talk at the Nobel symposium in Sweden on the concept of energy, and I wondered, Why did they invite me there? What am I going to tell these people who study solar energy, oil, wind? What can I tell them? And then I told them, "If you want to get lots of energy, you can start from practically nothing, and you can get all the energy in the universe."

Not everyone knows that when the universe expands, the total energy of matter does change. The total energy of matter plus gravity *does not* change, and it amounts to exactly zero. So the energy conservation for the universe is always satisfied, but it is trivial: Zero equals zero. But we are not interested in the energy of the universe as a whole; we are interested in the energy of matter.

If we can have a regime where we have some kind of instability, where the initial zero energy can split into a very big positive energy of matter and a very big negative energy of gravity, the total sum remains zero. But the total energy of matter can become as large as we want. This is one of the main ideas of inflation.

We have found how to start this instability, and how to stop it, because if it doesn't stop, then it goes on forever, and then it's not the universe where we can we live. Alan Guth's idea was how to start inflation, but he didn't know how to stop it in a graceful way. My idea was how to start it, continue it, and eventually stop it without damaging the universe. And when we learned how to do it, we understood that yes, we can start from practically nothing, or even literally nothing, as suggested by Alex Vilenkin, and account for everything that we see now. At that time, it was quite a revolutionary development: We finally could understand many properties of our universe. We no longer needed to postulate the cosmological principle; we finally knew the real physical reason why the world we see around us is uniform.

But then, soon after inventing new inflation, I realized something else. If you take a universe which initially was tiny, tiny, but still contained different parts with different properties, then our part of the universe might have exploded exponentially, and we no longer see other parts of the universe, which become far away from us because the distance between different parts of the universe increased exponentially during inflation. And those who live in other parts of the universe will be unable see us, because we will be far away. We look around, we do not see other parts of the universe with different properties, and so we think, "This is our universe, it is everywhere the same, other parts do not exist." And those who live in other parts of the universe will also think, "All of the universe is the way we see it."

For example, I could start in a red part of the universe, like in the Soviet Union, and you can start in a blue universe, and then after inflation, after each part becomes exponentially large, each of us would look around and say, just like Einstein and Newton did, "This is the universe, this is the whole thing, it is single-colored." And then some of us will try to explain why the universe must be red, and others will try to explain why the universe must be blue, all over the world. But now we know that from the point of view of inflation, it's quite possible that our universe is divided into many regions with different properties. Instead of the cosmological principle—which asserted that the whole world is the same everywhere and all of us must live in parts of the universe with similar properties—we are coming to a more cosmopolitan perspective: We live in a huge inflationary multiverse. Some of us can live in its red parts, some can live in blue parts, and there is nothing wrong about this picture as long as each of its parts is enormously large because of inflation.

Thus inflationary theory explained why our part of the universe looks so uniform: Everything that surrounds us was created by the exponential expansion of a tiny part of the universe. But we cannot see what happens at a distance much greater than the speed of light multiplied by the age of the universe. Inflationary theory tells that our part of the universe, the part we can see, is much, much smaller than the whole universe. Inflation of other parts of the universe may produce enormous regions with different properties. This was the first realization, which paved the way towards the theory of the inflationary multiverse. And the second realization was that even if we start with the same universe everywhere—a red universe, say— quantum fluctuations produced during inflation could make it multicolored.

I'm talking about different colors here only to help us visualize what happens during inflation and after it. Let me explain what I actually mean by that. Think about water. It can be liquid water, solid, or vapor. The chemical composition is the same, H_2O. But fish can live only in liquid water. Liquid, solid, or vapor are different phases of water. The same may happen with different realizations of the laws of physics in the universe. We usually assume, for simplicity, that all parts of the universe obey the same fundamental laws of physics. Nevertheless, different parts of the universe may dramatically differ from each other, just as icebergs differ from the water surrounding them. But instead of saying that water can be in different phases in application of physical laws, we say that different parts of the universe may be in different vacuum states, and in each of them the same fundamental law of physics may be realized differently. For example, in some parts of the universe we have weak, strong, and electromagnetic interactions, and in other parts these interactions do not differ from each other.

If we started in a red part of the universe, it does not stay red forever. Inflation is capable of producing and amplifying quantum fluctuations, and these quantum fluctuations let us jump from one vacuum state to another, and then to another. The universe becomes multicolored. The basic mechanism was understood as soon as the inflationary theory was invented. But its most interesting consequences appear in string theory.

A long time ago, string theorists realized that this theory allows many different vacua, so just by saying that we deal with a given version of string theory, one cannot actually predict the properties of our world, which depend on the choice of the vacuum state. Many people thought that this multiplicity of possible outcomes is a real problem with string theory. But in the context of the theory of inflationary multiverse, this is no longer a problem. In one of my papers, written in 1986, I said that it is actually a *virtue* of string theory. It allows creation of universes of many different types, with different laws of low-energy physics operating in each of them.

This simplifies the difficult task of explaining why the laws of physics in our part of the universe so nicely match the conditions required for our existence. Instead of the cosmological principle asserting that all parts of the world look alike, we found justification of the cosmological anthropic principle. It says that different exponentially large parts of the universe may be very different from each other, and we live only in those parts where life as we know it is possible. Thirty years ago, ideas of this type were extremely unpopular, but two decades later, when we learned more about properties of string theory vacua, the situation changed dramatically. The picture outlined above became a part of what Lenny Susskind called the string-theory landscape.

The way from the invention of inflationary theory to the string theory landscape has taken a very long time. In the beginning, it was pretty hard to work in this direction. Nobody wanted it. Nobody expected that inflationary theory, which was invented to explain the observed uniformity of the universe, would end up predicting that on a much greater scale than we can see now, the universe is 100-percent nonuniform. It was just too much. Even the simplest versions of inflationary theory seemed a little bit too revolutionary, providing a possibility of creating everything from practically nothing. It seemed like science fiction; it was too bold, too exotic.

For example, I found that in the first model of new inflation, which I invented back in 1981, the universe could expand 10^{800} times during the inflationary stage. It was surreal; we had never seen numbers like that in physics. When I was giving my first talks on new inflation at Lebedev Physical Institute, where I invented this theory, I had to apologize all the time, saying that 10^{800} was way too much. "Probably later," I said, "we'll come to something more realistic, the numbers will decrease and everything will become smaller." But then I invented a better inflationary theory, the theory of chaotic inflation, and the number became $10^{1000000000000}$. And then I found that inflation in this theory may continue eternally.

Inflationary theory explained many properties of our world, which we could not understand without it, and it also predicted many things that were confirmed experimentally. For example, quantum fluctuations produced during inflation, if the theory is right, explain formation of galaxies of the same type as our galaxy.

Andrei Linde

Secondly, there is the cosmic microwave background (CMB), which reaches our Earth from all parts of the sky. It has very interesting properties. When this radiation was discovered, almost fifty years ago, its temperature was measured, 2.7° Kelvin: very, very small. It comes to us from all parts of the universe in a uniform way. Then people looked at it more carefully, and they found that its temperature is 2.7° Kelvin plus 10^{-3}° Kelvin to the right of you, and 2.7° minus 10^{-3}° Kelvin to the left of you—a strange thing. But then an interpretation came. We are moving with respect to the universe, and there's a redshift. One part of the sky, in the direction of our motion, seems a little bit warmer, and another part of the sky seems a little bit cooler, because of the redshift, and that's OK.

Then people measured the temperature of the cosmic microwave radiation with an accuracy of about 10^{-5}° Kelvin. That's what the COBE satellite did, and that's what the WMAP satellite is doing right now, and many other experiments. And they have found many tiny spots in the sky—some a little bit warmer, some a little bit colder. They classified these spots, studied their distribution, and found that the distribution of spots was in a total agreement with the predictions of the simplest versions of inflationary theory. That was quite an unexpected confirmation. Theoretically, we knew that it should be there. What was incredible is that people who make observations could achieve this absolutely astonishing accuracy. It's just amazing.

Nobel Prizes are awarded for definitely established facts. Observers found fluctuations in the CMB, but one may wonder whether they were produced by inflation or by some other mechanism. So far, from my point of view, there are no other fully developed theories that would explain these observations. Who knows, maybe ten years from now, somebody will come up with something as smart as inflation. However, many people already tried to do it during the last thirty years, but no other theory comparable to inflation has been invented yet.

Let me tell you a little bit about how inflation was invented, from my perspective. Because things look very, very different when you look at it from a Russian perspective than from an American perspective, simply because at the time inflation theory was developed, Russia and America were not closely connected; every letter from Russia would come to the United States with an interval of about one-and-a-half or two months. You know, in the '30s there was the record flight from Russia landing in the United States, and it took almost a day, and then fifty years later a letter from Russia to the United States would take maybe forty to sixty days, so that was the progress achieved. Whatever.

I was educated at Moscow State University. I graduated, and then became a postdoc at Lebedev Physical Institute, where I worked with David Kirzhnits, an expert in condensed matter physics, in astrophysics, in quantum field theory, and nonlocal theories, and everything else. He was a really amazing person. He discovered the theory of cosmological phase transitions, and I further developed it. So one of the things that became a basis for many inflationary models was that in the early universe, physics could be completely different, and there may have been no difference between the weak, strong, and electromagnetic interactions.

Then for a while we thought that maybe this was just an exotic idea. I mean, not exotic—we knew it was true, but we were afraid it was untestable, like a Platonic idea, of which we would never know any consequences. But we were overly pessimistic. Some consequences of what we studied included the creation of strange objects in the early universe called primordial monopoles. If the theory we developed was right, and if unified theories of

Andrei Linde

weak, strong, and electromagnetic interactions proposed in the '70s were right—and at that time everybody thought they were right—then in the process of the cosmological phase transitions in the very early universe, when the whole world cooled down from its original hot state—which is what everybody believed it was—the theory predicted creation of some strange objects that look like separate south and north magnetic poles.

If you cut a magnet into two pieces, each of its parts will always have both a south pole and a north pole; you cannot have a south pole or a north pole separately. But according to these unified theories, there would also be objects in the universe called monopoles, each of them having either a south pole or a north pole. And each of them would be a million billion times heavier than the proton. People then calculated how many there should be, and the answer was that the total number of monopoles right now should be approximately the same as the number of protons, which would make our universe a million billion times heavier than it is now. Such a universe would be closed, and it would have collapsed already, in the first seconds of its evolution. Since we're still around, there's something wrong with this theory. So this was the primordial-monopole problem. It was a consequence of the theory of cosmological phase transitions we developed.

Many people were trying to resolve this problem, and they could not. Somehow, consistent solutions would not appear. Then, in parallel with these studies, we learned about a new cosmological theory proposed by Alexei Starobinsky in Russia, in '79 and '80. It was a rather exotic model where quantum corrections in quantum gravity produced the state of exponential expansion of the universe. This was a very interesting scenario, very similar to the inflation. But nobody called it inflationary theory, because it was proposed a year before Alan Guth proposed his "old inflation." It was called the Starobinsky model. It was the number-one subject of all discussions and debates at all cosmology conferences in Russia. But it didn't propagate to the United States, and one of the reasons it didn't was that the goal of Starobinsky was to solve the singularity problem and the model did not quite solve it, and he did not attempt to solve other problems, which were addressed by inflationary cosmology. In fact, he assumed that the universe was homogeneous from the very beginning; he did not try to explain why it was homogeneous. Nevertheless, when we look back at the history of inflationary models, formally, this model had all the features of successful inflationary models, except for the problem of the beginning of an inflationary stage. And the motivation for this [Starobinsky] model was rather obscure. But it worked, or almost worked.

Then Alan Guth formulated his model. But I didn't know about it. I attended a seminar at the Institute of Nuclear Research in Moscow, sometime in '80. At that seminar, Valery Rubakov, one of the famous Russian scientists who still lives and works there, discussed the possibility of solving the flatness and homogeneity problems by exponential expansion of the universe due to cosmological phase transitions in the so-called Coleman-Weinberg model. They were discussing all of these things without knowing anything about the paper by Alan Guth. They explained the problems, and they also explained why their model didn't solve these problems. Then they sent the paper for publication, but it was not accepted, because by that time the paper by Alan Guth had already come out.

Thus I learned about the idea of how one could solve many different cosmological problems during exponential expansion in the false vacuum, but I learned it not from Alan's paper. It happened quite a while before Alan's paper appeared in the Soviet Union, and the delay was just this mail delay. Moreover, I had known that one can have exponential expansion in the false vacuum since 1978, when we worked on it with Gennady Chibisov, but we found that this model didn't work, and so we didn't continue studying it; we didn't know that it might be useful for solving many cosmological problems until Rubakov and his collaborators told us about it.

When Alan's preprint arrived in Russia, Lev Okun, one of the famous Russian scientists, called me and asked: "You know, I heard something about the paper by Alan Guth, who is trying to solve the flatness problem by exponential expansion of the universe. Have you heard about it, anything?" And I told him, "No, I have not heard anything about it, but let me tell you how it works, and let me tell you why it doesn't work, in fact." So, for half an hour I explained to him Alan Guth's paper, although I did not see it at that time, and explained to him why it didn't work. And then, after a while, I received Alan Guth's paper. We had been in correspondence with Alan some time before that, to discuss the cosmological phase transitions, the expansion of the universe, and bubbles. But we did not discuss the possibility of using it for solving cosmological problems.

So that was it. If I had wanted to write a paper about that, I could have done it immediately after Okun called me, but it would have been completely dishonest, so I didn't even think about it, but of course I knew the basic ideas and I knew why they wouldn't work. So I didn't do anything. In fact, I was in a pretty depressed state, because the idea, obviously, was very beautiful and the idea that I'd learned from Rubakov was very, very beautiful. And it was a real pain that one could not make it work. The problems were extremely difficult.

A year later, Alan Guth, together with Erick Weinberg, wrote a seventy-page-long paper proving that it was impossible to improve Alan's model. Fortunately, I received it after I had already improved it—again, many thanks to the slow mail from the United States to Russia. While they were working on the paper, I was working on a solution. I found the solution somewhere in the summer of '81. In order to check whether this solution was obvious or not, I called Rubakov to check with him, because he had first introduced me to this set of ideas. In fact, my solution was so obvious that when it occurs to you, it seems simple. You cannot understand why you didn't think of it before.

I called him late in the evening. Because at that time my wife and kids were sleeping, I took the phone to our bathroom, and I was sitting on the floor there dialing him, checking with him, and he told me, "No, I haven't heard about it." So at that time I woke up my wife and I said, "You know, I think I know how the universe was created." And then I wrote—this was the summer of '81—I wrote a paper about it. But it took about three months to get permission for its publication.

At that time in Russia, if you wanted to send a paper in for publication abroad, first of all you had to do a lot of bureaucratic work in your institute. You typed it in Russian and got lots of signatures, then you sent it to the Academy of Sciences of the USSR, and they would send it to some other place that checked whether it was possible to publish it, whether it contained some important secrets which should remain that way. Then you received it back, typed it once in English, for a preprint, and typed it the second time, the same thing—no Xeroxes. And only then could you send it for publication. The whole procedure usually took two months, sometimes three months. I got permission only in October. But in October there was a conference in Moscow,

Andrei Linde

on quantum cosmology. And the best people came. Stephen Hawking came, many other good people came, and I gave a talk on this. And people were kind of excited, really excited about this, and they immediately offered their help to smuggle it from Russia as a preprint, and they would send it for publication, permitted or not. You know, friends, OK? Good friends can do it for you. But then there was an unexpected complication.

The morning after I gave the talk at this conference, I found myself at the talk. . . . Oh my god, this is going to be a funny story! I found myself at the talk by Stephen Hawking at the Sternberg Institute of Astronomy in Moscow University. I came there by chance, because I had heard from somebody that Hawking was giving a talk there. And they asked me to translate. I was surprised. OK, I will do it. Usually, at that time, Stephen would give his talk well prepared, which means his student would deliver the talk and Stephen from time to time would say something, and then the student would stop and change his presentation and say something else. So Stephen Hawking would correct and guide the student. But in this case they were completely unprepared. The talk was about inflation. The talk was about the impossibility of improving Alan Guth's inflationary theory.

So they were unprepared; they had just finished their own paper on it. As a result, Stephen would say one word, his student would say one word, and then they waited until Stephen would say another word, and I would translate this word. And all of these people in the auditorium, the best scientists in Russia, were waiting and asking: "What is going on? What is it all about?" So I decided let's just do it, because I knew what it was all about. So Stephen would say one word, the student would say one word, and then, after that, I would talk for five minutes, explaining what they were trying to say.

For about half an hour, we were talking this way and explaining to everyone why it was impossible to improve Alan Guth's inflationary model and what are the problems with it. And then Stephen said something and his student said, "Andrei Linde recently proposed a way to overcome this difficulty." I didn't expect that, and I happily translated it into Russian. And then Stephen said, "But this suggestion is wrong." And I translated it. . . . For half an hour, I was translating what Stephen said, explaining in great detail why what I'm doing is totally wrong. And it was all happening in front of the best physicists in Moscow, and my future in physics depended on them. I've never been in a more embarrassing situation in my life.

Then the talk was over, and I said, "I translated, but I disagree," and I explained why. And then I asked Stephen, "Would you like me to explain it to you in greater detail?" and he said, "Yeah." And then he rode out from this place and we found some room, and for about two or three hours all the people in Sternberg Institute were in a panic, because the famous British scientist had just disappeared, nobody knew where to.

During that time, I was near the blackboard, explaining what was going on there. From time to time, Stephen would say something, and his student would translate: "But you didn't say that before." Then I would continue, and Stephen would again say something, and his student would say again the same words, "But you didn't say that before." And after we finished, I jumped into his car and they brought me to their hotel. We continued the discussion, which ended by him showing me photographs of his family, and we became friends. He later invited me to a conference in Cambridge, England, which was specifically dedicated to inflationary theory. So that's how it all started. It was pretty dramatic.

Andrei Linde

Because Stephen had made some objections to what I said, I polished the end of the paper to respond to some of his objections. I didn't give it to my friends to smuggle from Russia. So I sent it for publication in October. It arrived eventually at *Physics Letters*, but because it was delayed, it was published in '82 instead of '81. And I also sent lots of preprints to the United States, and one of them reached Paul Steinhardt and Andy Albrecht, who both worked on similar ideas. Three months after I sent my paper for publication, they sent for publication their own paper, with the same idea described in it, and with a reference to my work.

It was a miracle that the government allowed me to go to Cambridge. I had visited Italy previously, but then for a while, for some reasons which were not explained to me, they were unwilling to let me go anywhere outside the Soviet Union. But that time it worked, and it was the most wonderful conference in my life. It was the first conference on inflationary cosmology attended by the best people in this area. Things really *happened* at this conference. It was magnificent. Three weeks of intense discussion and work together.

The conference was in summer 1982. The whole conference was about new inflation. I gave this name to the scenario I developed. But this theory, just like the old scenario proposed by Guth, did not live long. Because of this symposium, new inflation in its original form essentially died in '82. The theory predicted too large perturbations of density. The model required modifications, and these modifications were such that there could be no thermal equilibrium in the universe, no cosmological phase transitions, so no way to realize a scenario like Alan Guth and I envisioned. Interestingly, most of the books on astronomy still describe inflation as exponential expansion during the cosmological phase transitions; this theory was so popular that nobody even noticed that it died back in '82. But a year later, in '83, I invented a different scenario, which was actually much simpler. It was chaotic inflation, and it did not require the universe to be hot to start with.

In the chaotic-inflation scenario, one could have an inflationary regime without assuming that the universe initially was hot. I abandoned the idea of the cosmological phase transitions, metastability, false vacua—most of the things that formed the basis for the old inflation model proposed by Guth and for my own new-inflationary scenario. After all of these modifications, the inflationary regime became much simpler, more general, and it could exist in a much broader class of theories. In '86 I found that if we have inflation in the simplest chaotic inflation models, then, because of quantum fluctuations, inflation would go on forever in some parts of the universe. Alex Vilenkin found a similar effect for the new-inflation scenario. The effect that I found was very generic. I called it eternal inflation.

What Vilenkin studied was the theory of new inflation, and in new inflation, you can start at the top of the potential energy and the field doesn't know whether to roll down to the right or roll down to the left, so while you stay at the top you're thinking you'll fall down, but you can think for a long time, and during this time the expansion of the universe produces lots of volume. In chaotic inflation, where the potential energy has the simplest parabolic form—no specifically flat pieces of potential are required—you just take a model like that, and if the field is sufficiently high, there are quantum fluctuations, and the scalar field wants to go down, but quantum fluctuations sometimes throw it higher. The probability of jumping high is very small, but if you jump, you are exponentially rewarded by the creation of huge amounts of new volume of the universe. You start with a

Andrei Linde

tiny part of the universe, and then it just spreads and spreads. It's like a chain reaction. It's called the branching diffusion process.

This is the basic idea of eternal inflation. In the paper of '86, where I discovered eternal chaotic inflation, I also noted that if you have eternal inflation in string theory, then the universe will be divided into an enormous number of different, exponentially large parts with different properties corresponding to a large number of different stringy vacua, and that's an advantage. That was what later became the string theory landscape.

I should say that one of the most important predictions of inflation was the theory of quantum fluctuations, which give rise to galaxies eventually. Just think about it. If inflation were not to produce inhomogeneities, then when it blows up and the universe becomes almost exactly homogeneous, that would be the end of the game: no galaxies, no life. We would be unable to live in an exactly uniform universe, because it would be empty.

Fortunately, there was a way around it. Before I even introduced new inflation, I knew about the interesting work of two men from Lebedev Physical Institute, Gennady Chibisov and Slava Mukhanov. Mukhanov was younger, but he was the ideological leader in this group. By studying the Starobinsky model, which, as we now know, is a version of inflationary theory, they found that during inflation in the Starobinsky model, quantum fluctuations grow and may eventually give rise to galaxies. And we looked at them and said, "Oh, come on, guys. You cannot be right. It's impossible, because galaxies are big classical objects and you are starting with nothing. You start with quantum fluctuations."

What they managed to explain to us—and this was an important ingredient that influenced everything we did later—was that these quantum fluctuations become essentially classical when the universe becomes large. They give rise to galaxy formation. Their paper of 1981 was the first paper on that subject. After that, in '82, similar ideas were rediscovered by a group of people who were all at the Cambridge symposium. This group included Hawking, Starobinsky, Guth, Bardeen, Steinhardt, and Turner. Their ideas were developed in application to new inflation, but it all started with Chibisov and Mukhanov in '81. And then Mukhanov continued studying it more and more, and he developed a general theory of these quantum fluctuations. From my perspective, this is one of the most important parts of inflationary theory. This is most important not only for the theory of galaxy formation. Chaotic eternal inflation would be impossible if not for these quantum fluctuations.

Some of the most interesting recent developments of inflationary cosmology are related to string theory. My understanding of this theory is based in part on my collaboration with my wife, Renata Kallosh. She is also a professor of physics at Stanford; she studies supergravity and string theory. So let me tell you about these theories just a little.

During the last years of his life, Einstein dreamed about a final theory which would unify symmetries of space with symmetries of elementary particles. And he failed. I was told that during the last years of his life he continued writing on the blackboard, filling it with equations of the new theory, and although he could not successfully finish it, he was still happy. Then people learned there was a no-go theorem. One just couldn't do it, period. It was impossible to realize Einstein's dream of a unified theory of everything. Then other people found that there was a loophole in the no-go theorem. If the theory has a special symmetry, supersymmetry, relating to each other bosons (scalar fields, photons) and fermions (quarks and leptons), then these no-go theorems go away.

Andrei Linde

Thus, it all began with supersymmetry, and then it became a more advanced theory: supergravity. One could unify the theory of gravity with the theory of elementary particles. It was fantastic! The theory flourished in the middle of the '70s up to the '80s. It resolved some problems of quantum gravity. Some infinite expressions, which appeared in calculations in a quantum theory of gravity, disappeared in supergravity. Everybody was ecstatic, until the moment they found that these infinities might still appear in supergravity in the third approximation, or maybe in the eighth approximation. Something was not quite working—although some very recent results suggest that maybe people were too pessimistic at that time and some versions of the theory of supergravity are quite good.

But at that time, they looked at it and said, "OK, it doesn't work, there are some problems with the theory, can we do something about it?" The next step was the theory of superstrings. The development of science was not like, "Oh, come on, we can go to the right, we can go to the left, we can go anywhere, let's go straight." No, it was not like that. We wanted to achieve the unity of all forces, and because of the no-go theorem there was no way to do it except by using supersymmetry. Then it becomes supergravity. You just must have it, if you want to describe curved space in supersymmetric theories. Now you have supergravity, but . . . sorry, it doesn't quite work, you need to somehow generalize it. And then string theory was developed.

This was like a valley in the mountains. It's not about going to the right or to the left. Your valley shows you the best way, maybe even the only way. That's how people came to string theory, and then they became very optimistic. This was '85. They were thinking they would do everything pretty quickly. I must say that not everybody was so optimistic at that time. In particular, John Schwarz, one of the fathers of string theory, said, "Oh, well, it may actually take more than twenty years for string theory to come to fruition as a phenomenological theory of everything." He made a warning. Well, enthusiasm was nevertheless overwhelming, which was good and bad. It was good because so many talented young people entered the field. It was bad because the supergravity tradition was partially forgotten. In Europe, the supergravity tradition is still alive, very much so. In the United States, it's not that much.

String theory is based on the idea that our universe fundamentally has more dimensions, not just four. This idea was also part of some versions of supergravity. It was also a part of Kaluza-Klein theory, a long time ago. The standard attitude was that string theory required an assumption that our space is ten-dimensional and six dimensions should be compactified. After that, we have three large dimensions of space and one of time. The other six dimensions would be very small. Superstring people often use Calabi-Yau space to describe compactification of the six extra dimensions; this space can have a very complicated topology.

The question, though, was "How do we know that this is true?" For a long time, nobody could construct a working mechanism that would allow Calabi-Yau space to be really small. Why do we need it to be small? Because we cannot move in these six dimensions, we are too big for that. We can go to the right, to the left, and upward, but we cannot go in six other directions. At least, nobody told me that they tried, had been there.

Six dimensions. We needed to explain why they're small. There was a property, an unfortunate property, of string theory that if treated naïvely, without any special effort, these six dimensions actually want to decompactify—want to spread out, become large. There could be many ways of compactifying space, which is the origin of many different vacua in string theory. But nevertheless the problem was how to keep these extra dimensions small.

The attitude was, "Oh, well, we'll do something. We will do something." But this "we will do something" continued for almost twenty years. Nobody took this problem too seriously, because there were lots of other problems in string theory and they thought, "Let's just go forward." But we needed to study these string vacua. "The vacuum" means the state that looks empty from our four-dimensional point of view, but its properties depend on the properties of the compactified Calabi-Yau space, the compactified six-dimensional space. The vacuum does not contain particles. If we add the particles, then we can have our universe. This vacuum, this place without particles, galaxies, us—what properties does this vacuum have? As I said, in order to study it consistently, we need to have stable compactification of the extra six dimensions of space. There are also other fields in this theory that need to be stabilized. People didn't know how to do it, but for a while it was not such an urgent problem. But then, at the end of the '90s, cosmologists discovered the exponential, accelerating expansion of our universe, which happens because of what people call dark energy—or the cosmological constant. This discovery made a very strong impact on the development of string theory.

The rumor was that at a conference in India, Ed Witten, who is the leading authority in string theory, said that he didn't actually know how to explain the acceleration of the universe in the context of string theory. And when Ed Witten doesn't know something, people start taking it seriously and panicking a bit. So at that moment they started really paying attention to the properties of the vacuum in string theory. They wanted to explain this exponential expansion of the universe, which apparently started about 5 billion years ago and goes on very slowly. So people started trying to explain it, and it didn't work. Then they started thinking even more attentively about this problem—what actually defines this vacuum state, what stabilizes it.

In 2003, I was part of a group at Stanford—Kallosh, Kachru, Linde, and Trivedi—that proposed a possible solution of this problem. There were some earlier works, I must say, which came very close to a solution, and now there are some other ways of doing it. But this was kind of like a point of crystallization, where people realized that we could actually solve the problem and stabilize the vacuum in string theory.

When we found a way to do it, it was immediately realized that there are exponentially *many* ways to do it. People who estimated the total number of different ways to stabilize the vacuum in string theory came up with astonishing numbers, like 10^{500}. Michael Douglas and his collaborators made this estimate. And this fact has profound cosmological implications. If you marry string theory with the theory of eternal inflation, then you can have one type of vacuum in one part of the universe, another vacuum in another part of the universe, and it is possible to jump from one vacuum to another, due to quantum effects. Lenny Susskind gave this scenario a very catchy name, the string theory landscape.

What I mean is that when we're talking about this vacuum state, "vacuum state" means the homogeneous state describing our three dimensions—three dimensions plus one [of time]. But the remaining six dimensions, they may squeeze like this or they may squeeze like that. There are lots of different topologies. In addition to different topologies, there are different fields that can exist in this six-dimensional space—so-called fluxes.

Andrei Linde

There are other objects that can exist there and that can determine properties of our space. In our space we do not see them; they are in this tiny six-dimensional compactified space. But they determine properties of our vacuum—in particular, the vacuum energy density. The level of this vacuum energy depends on what is going on in the compactified space. Properties of elementary-particle physics depend on what happens there. If you have many different ways of compactification, you have the same string theory fundamentally, but your world—your three-dimensional space and one dimension of time—will have completely different properties. That's what is called the string theory landscape. You have the same string theory, but you have many different realizations of it. That is exactly what I envisioned in my paper on eternal chaotic inflation in '86: We have lots of possibilities, and this is good.

But in '86 we didn't know a single example of a stable string theory vacuum; we just expected that there should be exponentially many such vacua. In 2003, we learned how to find such vacua, and then it was realized that indeed there are lots and lots of them. So that is the present view.

Let me say a few words about what I'm studying right now. 10^{500} is an abnormally large number; it tells you how many choices of vacua you have. You have this huge amount of possibilities. And by the way, there's a question which many people ask: "How do you know?" How do we know that we have this multitude—that these other parts of the universe are somewhere inside our universe?

This is the picture: The universe is very, very big, and it's divided into parts. Here is one realization of the string of vacua. There, in the same universe, but far away from us, it's a different vacuum. The guys here and there don't know about each other, because they're exponentially far apart. That's important to understand in order to have a vision of the universe. It's important that you have a choice. But if you don't see these parts, how do you know they actually exist, and why do you care?

Usually I answer in the following way: If we do not have this picture, then we cannot explain the many strange coincidences that occur around us. Like why the vacuum energy is so immensely small, incredibly small. Well, that's because we have many different vacua, and in those vacua where the vacuum energy is too large, galaxies cannot form. In those vacua where the energy density is negative, the universe rapidly collapses. And in our vacuum the energy density is just right, and that's why we live here. That's the anthropic principle. But you cannot use the anthropic principle if you don't have many possibilities to choose from. That's why the multiverse is so desirable, and that's what I consider experimental evidence in favor of the multiverse.

I introduced the anthropic principle in the context of the inflationary multiverse back in '82. The idea of new inflation was proposed in '81, and then in '82 I wrote two papers where I emphasized the anthropic principle in the context of inflationary cosmology. I said that the universe may consist of many different exponentially large parts. I did not use the word "multiverse," I just said that the universe may consist of many, many mini-universes with different properties, and I've studied this possibility since that time, for many, many years.

But what is important is that when we studied inflationary theory, we started asking questions that seemed to be metaphysical, like why parallel lines do not intersect, why the universe is so big. And if we had said, "Oh my god, these are metaphysical questions and we should not venture into it," then we would never have discovered the solutions. Now we're asking metaphysical questions about the anthropic principle, about stuff like that, and many, many people tell us, "Don't do it, this is bad, this is the 'a' word. You should avoid it."

We shouldn't avoid anything. We should try to do our best to use the simplest explanations possible, or what proves simplest, and if something falls into your hands as an explanation of why the cosmological constant vacuum energy is so small, and you decide not to accept it for ideological reasons, this is very much what we had in Russia long ago. That ideology told me which type of physics was right and which type of physics was wrong. We should not proceed this way. Once you have multiple possibilities, then you can have scientific premises for anthropic considerations, not just philosophical talk about "other worlds." Now we have a consistent picture of the multiverse, so now we can say, "This is physics. This is something serious."

When you look at our own part of the universe, you have a galaxy to the right of you, you have a set of galaxies to the left of you. Could it be that our universe was formed differently? Is there a chance that me, my copy, might live somewhere far away from me? How far away from me? Why my exact copy? Well, because quantum fluctuations produce different universes over and over again. Alex Vilenkin has this description of "many worlds in one." He wrote a book about it. The question is, how many of these different types of universes can you produce? And "different types of universes" does not just mean vacuum states but different distributions of matter. The distribution of matter in our part of the universe, the distribution of galaxies, is determined by quantum fluctuations, which were produced during inflation.

For example, you may have a scalar field that rolls down slowly, which is how inflation ends, but then quantum fluctuations push it, locally, up just a little. Then in these places the energy of the universe will locally increase. Then it becomes very, very big, and at this part you'll see a galaxy in the place where you live. If the jump is down, then in this place you will see no galaxies. When this happens during inflation over and over again at different scales, then you check: How many different jumps, or kinds of configurations of jumps, have there been?

These jumps produce what later looks like different classical universes—a galaxy here, a void there. How many possibilities are there? And the answer—and this is purely a combinatorial answer—is that if n is the number of times the size of the universe doubled during inflation, and you take 2^{3n}, this will show you the volume of the universe after inflation. Where the volume grows by 2^{3n}, the total number of possible configurations that can occur there because of these quantum jumps will be also proportional to 2^{3n}. This will give you the total number of possible configurations of matter that you can produce during inflation, and this number typically is much, much greater than 10^{500}.

Of course, during eternal inflation, inflation goes on forever, so you could even expect that this number is infinite. However, during eternal inflation each jump can be repeated; it can repeat itself. A scalar field jumps again to the state where it jumps again, to a state where it jumps again, and eventually it starts producing identical configurations of matter.

Think about it this way: Previously we thought our universe was like a spherical balloon. In the new picture, it's like a balloon producing balloons producing balloons. This is a big fractal. The Greeks thought of our universe as an ideal sphere because this was the best image they had at their disposal. The 20th-century idea is a fractal, the beauty of a fractal. Now you have these fractals. We ask, How many different types of these elements of fractals are there which are irreducible to each other? And the number will be exponentially large. In the simplest models, it's about 10 to the degree 10, to the degree 10, to the degree 7. It actually may be much more than that, even though nobody can see all of these universes at once.

Soon after Alan Guth proposed his version of inflationary theory, he famously exclaimed that the universe is an ultimate free lunch. Indeed, in inflationary theory the whole universe emerges from almost nothing. A year later, in the proceedings of the first conference on inflation, in Cambridge, I expanded his statement by saying that the universe is not just a free lunch, it is an eternal feast where all possible dishes are served. But at that time I could not even imagine that the menu of all possible universes could be so incredibly large.

5
Theories of the Brane

Lisa Randall

Theoretical physicist, Frank B. Baird, Jr., Professor of Science, Harvard University; author, *Higgs Discovery: The Power of Empty Space*

Particle physics has contributed to our understanding of many phenomena, ranging from the inner workings of the proton to the evolution of the observed universe. Nonetheless, fundamental questions remain unresolved, motivating speculations beyond what is already known. These mysteries include the perplexing masses of elementary particles; the nature of the dark matter and dark energy that constitute the bulk of the universe; and what predictions string theory, the best candidate for a theory incorporating both quantum mechanics and general relativity, makes about our observed world. Such questions, along with basic curiosity, have prompted my excursions into theories that might underlie currently established knowledge. Some of my most recent work has been on the physics of extra dimensions of space and has proved rewarding beyond expectation.

Particle physics addresses questions about the forces we understand—the electromagnetic force, the weak forces associated with nuclear decay, and the strong force that binds quarks together into protons and neutrons—but we still have to understand how gravity fits into the picture. String theory is the leading contender, but we don't yet know how string theory reproduces all the particles and physical laws we actually see. How do we go from this pristine, beautiful theory existing in ten dimensions to the world surrounding us, which has only four—three spatial dimensions plus time? What has become of string theory's superfluous particles and dimensions?

Sometimes a fruitful approach to the big, seemingly intractable problems is to ask questions whose possible answers will be subject to experimental test. These questions generally address physical laws and processes we've already seen. Any new insights will almost certainly have implications for even more fundamental questions. For example, we still don't know what gives rise to the masses of the fundamental particles—the quarks, leptons (the electron, for example), and electroweak gauge bosons—or why these masses are so much less than the mass associated with quantum gravity. The discrepancy is not small: The two mass scales are separated by 16 orders of magnitude! Only theories that explain this huge ratio are likely candidates for theories underlying the standard model. We don't yet know what that theory is, but much of current particle physics research, including that involving extra dimensions of space, attempts to discover it. Such speculations will soon be explored at the Large Hadron Collider in Geneva, which will operate at the TeV energies relevant to particle physics. The results of experiments to be performed there should select among the various proposals for the underlying physical description in concrete and immediate ways. If the underlying theory turns out to be either supersymmetry or one of the extra-dimension theories I will go on to describe, it will have deep and lasting implications for our conception of the universe.

Right now, I'm investigating the physics of the TeV scale. Particle physicists measure energy in units of electron volts. "TeV" means "a trillion electron volts." This is a very high energy and challenges the limits of current technology, but it's low from the perspective of quantum gravity, whose consequences are likely to show up only at energies 16 orders of magnitude higher. This energy scale is interesting, because we know that the as-yet-undiscovered part of the theory associated with giving elementary particles their masses should be found there.

Most of us, however, suspect that a prerequisite for progress will be a worked-out theory that relates gravity to the microworld. Back at the very beginning, the entire universe could have been squeezed to the size of an elementary particle. Quantum fluctuations could shake the entire universe, and there would be an essential link between cosmology and the microworld. Of course, string theory and M-theory are the most ambitious and currently fashionable attempts to do that. When we have that theory, we at least ought to be able to formulate some physics for the very beginning of the universe. One question, of course, is whether we'll find that space and time are so complicated and screwed up that we can't really talk about a beginning in time. We've got to accept that we will have to jettison more and more of our commonsense concepts as we go to these extreme conditions.

The main stumbling block at the moment is that the mathematics involved in these theories is so difficult that it's not possible to relate the complexity of this ten- or eleven-dimensional space to anything we can actually observe. In addition, although these theories may appear aesthetically attractive, and although they give us a natural interpretation of gravity, they don't yet tell us why our three-dimensional world contains the types of particles that physicists study. We hope that one day this theory, which already deepens our insight into gravity, will gain credibility by explaining some of the features of the microworld that the current standard model of particle physics does not.

Lisa Randall

Although Roger Penrose can probably manage four dimensions, I don't think any of these theorists can in any intuitive way imagine the extra dimensions. They can, however, envision them as mathematical constructs, and certainly the mathematics can be written down and studied. The one thing that's rather unusual about string theory from the viewpoint of the sociology and history of science is that it's one of the few instances where physics has been held up by a lack of the relevant mathematics. In the past, physicists have generally taken fairly old-fashioned mathematics off the shelf. Einstein used 19th-century non-Euclidean geometry, and the pioneers in quantum theory used group theory and differential equations that had essentially been worked out long beforehand. But string theory poses mathematical problems that aren't yet solved, and has actually brought math and physics closer together.

String theory is the dominant approach right now, and it has some successes already, but the question is whether it will develop to the stage where we can actually solve problems that can be tested observationally. If we can't bridge the gap between this ten-dimensional theory and anything that we can observe, it will grind to a halt. In most versions of string theory, the extra dimensions above the normal three are all wrapped up very tightly, so that each point in our ordinary space is like a tightly wrapped origami in six dimensions. We see just three dimensions; the rest are invisible to us because they are wrapped up very tightly. If you look at a needle, it looks like a one-dimensional line from a long distance, but really it's three-dimensional. Likewise, the extra dimensions could be seen if you looked at things very closely. Space on a very tiny scale is grainy and complicated—its smoothness is an illusion of the large scale. That's the conventional view in these string theories.

An idea which has become popular in the last two or three years is that not all the extra dimensions are wrapped up—that there might be at least one extra dimension that exists on a large scale. Raman Sundrum and I have developed this idea in our work on branes. According to this theory, there could be other universes, perhaps separated from ours by just a microscopic distance; however, that distance is measured in some fourth spatial dimension, of which we are not aware. Because we are imprisoned in our three dimensions, we can't directly detect these other universes. It's rather like a whole lot of bugs crawling around on a big two-dimensional sheet of paper, who would be unaware of another set of bugs that might be crawling around on another sheet of paper that could be only a short distance away in the third dimension. In a different way, this concept features in a rather neat model that Paul Steinhardt and Neil Turok have discussed, which allows a perpetual and cyclic universe. These ideas, again, may lead to new insights. They make some not-yet-testable predictions about the fluctuation of gravitational waves, but the key question is whether they have the ring of truth about them. We may know that when they've been developed in more detail.

Two of the potential explanations for the huge disparity in energy scales are supersymmetry and the physics of extra dimensions. Supersymmetry, until very recently, was thought to be the only way to explain physics at the TeV scale. It is a symmetry that relates the properties of bosons to those of their partner fermions—bosons and fermions being two types of particles distinguished by quantum mechanics. Bosons have integral spin and fermions have half-integral spin, where spin is an internal quantum number. Without supersymmetry, one would expect these two particle types to be unrelated. But given supersymmetry, properties like mass and the interaction strength between

Lisa Randall

a particle and its supersymmetric partner are closely aligned. It would imply for an electron, for example, the existence of a corresponding superparticle—called a selectron in this case—with the same mass and charge.

There was and still is a big hope that we will find signatures of supersymmetry in the next generation of colliders. The discovery of supersymmetry would be a stunning achievement. It would be the first extension of symmetries associated with space and time since Einstein constructed his theory of general relativity in the early 20th century. And if supersymmetry is right, it's likely to solve other mysteries, such as the existence of dark matter. String theories that have the potential to encompass the standard model seem to require supersymmetry, so the search for supersymmetry is also important to string theorists. Both for these theoretical reasons and for its potential experimental testability, supersymmetry is a very exciting theory.

However, like many theories, supersymmetry looks fine in the abstract but leaves many questions unresolved when you get down to the concrete details of how it connects to the world we actually see. At some energy, supersymmetry must break down, because we haven't yet seen any "superpartners." This means that the two particle partners—for example, the electron and the selectron—cannot have exactly the same mass; if they did, we would see both. The unseen partner must have a bigger mass if it has so far eluded detection. We want to know how this could happen in a way consistent with all known properties of elementary particles. The problem for most theories incorporating supersymmetry breaking is that all sorts of other interactions and decays are predicted which experiment has already ruled out. The most obvious candidates for breaking supersymmetry permit the various kinds of quarks to mix together, and particles would have a poorly defined identity. The absence of this mixing, and the retention of the various quark identities, is a stringent constraint on the content of the physical theory associated with supersymmetry breaking, and is one important reason that people were not completely satisfied with supersymmetry as an explanation of the TeV scale. To find a consistent theory of supersymmetry requires introducing physics that gives masses to the supersymmetric partners of all the particles we know to exist, without introducing interactions we don't want. So it's reasonable to look around for other theories that might explain why particle masses are associated with the TeV energy scale and not one that's 16 orders of magnitude higher.

There was a lot of excitement when it was first suggested that extra dimensions provide alternative ways to address the origin of the TeV energy scale. Additional spatial dimensions may seem like a wild and crazy idea at first, but there are powerful reasons to believe that there really are extra dimensions of space. One reason resides in string theory, in which it's postulated that the particles are not themselves fundamental but are oscillation modes of a fundamental string. The consistent incorporation of quantum gravity is the major victory of string theory. But string theory also requires nine spatial dimensions, which, in our observable universe, is obviously six too many. The question of what happened to the six unseen dimensions is an important issue in string theory. But if you're coming at it from the point of view of the relatively low-energy questions, you can also ask whether extra dimensions could have interesting implications in our observable particle physics or in the particle physics that should be observable in the near future. Can extra dimensions help answer some of the unsolved problems of three-dimensional particle physics?

People entertained the idea of extra dimensions before string theory came along, although such speculations were soon forgotten or ignored. It's natural to ask what would happen if there were different dimensions of space; after all, the fact that we see only three spatial dimensions doesn't necessarily mean that only three exist, and Einstein's general relativity doesn't treat a three-dimensional universe preferentially. There could be many unseen ingredients to the universe. However, it was first believed that if additional dimensions existed they would have to be very small in order to have escaped our notice. The standard supposition in string theory was that the extra dimensions were curled up into incredibly tiny scales—10^{-33} centimeters, the so-called Planck length and the scale associated with quantum effects becoming relevant. In that sense, this scale is the obvious candidate: If there are extra dimensions, which are obviously important to gravitational structure, they'd be characterized by this particular distance scale. But if so, there would be very few implications for our world. Such dimensions would have no impact whatsoever on anything we see or experience.

From an experimental point of view, though, you can ask whether extra dimensions really must be this ridiculously small. How large could they be and still have escaped our notice? Without any new assumptions, it turns out that extra dimensions could be about 17 orders of magnitude larger than 10^{-33} cm. To understand this limit requires more fully understanding the implications of extra dimensions for particle physics.

If there are extra dimensions, the messengers that potentially herald their existence are particles known as Kaluza-Klein modes. These KK particles have the same charges as the particles we know, but they have momentum in the extra dimensions. They would thus appear to us as heavy particles with a characteristic mass spectrum determined by the extra dimensions' size and shape. Each particle we know of would have these KK partners, and we would expect to find them if the extra dimensions were large. The fact that we have not yet seen KK particles in the energy regimes we've explored experimentally puts a bound on the extra dimensions' size. As I mentioned, the TeV energy scale of 10^{-16} cm has been explored experimentally. Since we haven't yet seen KK modes and 10^{-16} cm would yield KK particles of about a TeV in mass, that means all sizes up to 10^{-16} are permissible for the possible extra dimensions. That's a lot larger than 10^{-33} cm, but it's still too small to be significant.

This is how things stood in the world of extra dimensions until very recently. It was thought that extra dimensions might be present but that they would be extremely small. But our expectations changed dramatically after 1995, when Joe Polchinski, of the University of California at Santa Barbara, and other theorists recognized the importance of additional objects in string theory called branes. Branes are essentially membranes—lower-dimensional objects in a higher-dimensional space. (To picture this, think of a shower curtain, virtually a two-dimensional object in a three-dimensional space.) Branes are special, particularly in the context of string theory, because there's a natural mechanism to confine particles to the brane; thus not everything need travel in the extra dimensions, even if those dimensions exist. Particles confined to the brane would have momentum and motion only along the brane, like water spots on the surface of your shower curtain.

Branes allow for an entirely new set of possibilities in the physics of extra dimensions, because particles confined to the brane would look more or less as they would in a three-plus-one-dimension world; they never venture beyond it. Protons, elec-

Lisa Randall

trons, quarks, all sorts of fundamental particles could be stuck on the brane. In that case, you may wonder why we should care about extra dimensions at all, since despite their existence the particles that make up our world do not traverse them. However, although all known standard-model particles stick to the brane, this is not true of gravity. The mechanisms for confining particles and forces mediated by the photon or electrogauge proton to the brane do not apply to gravity. Gravity, according to the theory of general relativity, must necessarily exist in the full geometry of space. Furthermore, a consistent gravitational theory requires that the graviton, the particle that mediates gravity, has to couple to any source of energy, whether that source is confined to the brane or not. Therefore, the graviton would also have to be out there in the region encompassing the full geometry of higher dimensions—a region known as the bulk—because there might be sources of energy there. Finally, there's a string-theory explanation of why the graviton is not stuck to any brane: The graviton is associated with the closed string, and only open strings can be anchored to a brane.

A scenario in which particles are confined to a brane and only gravity is sensitive to the additional dimensions permits extra dimensions that are considerably larger than previously thought. The reason is that gravity is not nearly as well tested as other forces, and if it's only gravity that experiences extra dimensions, the constraints are much more permissive. We haven't studied gravity as well as we've studied most other particles, because it's an extremely weak force and therefore more difficult to precisely test. Physicists have showed that even dimensions almost as big as a millimeter would be permitted if it were only gravity out in the higher-dimensional bulk. This size is huge compared with the scales we've been talking about. It is a macroscopic, visible size! But because photons (which we see with) are stuck to the brane, too, the dimensions would not be visible to us, at least in the conventional ways.

Once branes are included in the picture, you can start talking about crazily large extra dimensions. If the extra dimensions are very large, that might explain why gravity is so weak. Gravity might not seem weak to you, but it's the entire Earth that's pulling you down; the result of coupling an individual graviton to an individual particle is quite small. From the point of view of particle physics, which looks at the interactions of individual particles, gravity is an extremely weak force. This weakness of gravity is a reformulation of the so-called hierarchy problem— that is, why the huge Planck mass suppressing gravitational interactions is 16 orders of magnitude bigger than the mass associated with particles we see. But if gravity is spread out over large extra dimensions, its force would indeed be diluted. The gravitational field would spread out in the extra dimensions and consequently be very weak on the brane—an idea recently proposed by theorists Nima Arkani Hamed, Savas Dimopoulos, and Gia Dvali. The problem with this scenario is the difficulty of explaining why the dimensions should be so large. The problem of the large ratio of masses is transmuted into the problem of the large size of curled-up dimensions.

Raman Sundrum, currently at Johns Hopkins University, and I recognized that a more natural explanation for the weakness of gravity could be the direct result of the gravitational attraction associated with the brane itself. In addition to trapping particles, branes carry energy. We showed that from the perspective of general relativity this means that the brane curves the space around it, changing gravity in its vicinity. When the energy in space is correlated with the energy on the brane so that a large flat three-dimensional brane sits in the higher-dimensional space, the graviton—the particle communicating the gravitational force—is highly attracted to the brane. Rather than spreading uniformly in an extra dimension, gravity stays localized, very close to the brane.

The high concentration of the graviton near the brane—let's call the brane where gravity is localized the Planck brane—leads to a natural solution to the hierarchy problem in a universe with two branes. For the particular geometry that solves Einstein's equations, when you go out some distance in an extra dimension, you see an exponentially suppressed gravitational force. This is remarkable, because it means that a huge separation of mass scales—16 orders of magnitude—can result from a relatively modest separation of branes. If we are living on the second brane—not the Planck brane—we would find that gravity was very weak. Such a moderate distance between branes is not difficult to achieve and is many orders of magnitude smaller than that necessary for the large-extra-dimensions scenario just discussed. A localized graviton plus a second brane separated from the brane on which the standard model of particle physics is housed provides a natural solution to the hierarchy problem—the problem of why gravity is so incredibly weak. The strength of gravity depends on location, and away from the Planck brane it is exponentially suppressed.

This theory has exciting experimental implications, since it applies to a particle physics scale—namely, the TeV scale. In this theory's highly curved geometry, Kaluza-Klein particles—those particles with momentum in the extra dimensions—would have mass of about a TeV; thus there is a real possibility of producing them at colliders in the near future. They would be created like any other particle and they would decay in much the same way. Experiments could then look at their decay products and re-construct the mass and spin that is their distinguishing property. The graviton is the only particle we know about that has spin-2. The many Kaluza-Klein particles associated with the graviton would also have spin-2 and could therefore be readily identified. Observation of these particles would be strong evidence of the existence of additional dimensions and would suggest that this theory is correct.

As exciting as this explanation of the existence of very differ-ent mass scales is, Raman and I discovered something perhaps even more surprising. Conventionally, it was thought that extra dimensions must be curled up or bounded between two branes, or else we would observe higher-dimensional gravity. The afore-mentioned second brane appeared to serve two purposes: It ex-plained the hierarchy problem because of the small probability for the graviton to be there, and it was also responsible for bounding the extra dimension so that at long distances, bigger than the di-mension's size, only three dimensions are seen.

The concentration of the graviton near the Planck brane can, however, have an entirely different implication. If we forget the hierarchy problem for the moment, the second brane is unneces-sary. That is, even if there's an infinite extra dimension and we live on the Planck brane in this infinite dimension, we wouldn't know about it. In this "warped geometry," as the space with ex-ponentially decreasing graviton amplitude is known, we would see things as if this dimension did not exist and the world were only three-dimensional.

Because the graviton has such a small probability of being lo-cated away from the Planck brane, anything going on far away from the Planck brane should be irrelevant to physics on or near it. The physics far away is in fact so entirely irrelevant that the

Lisa Randall

extra dimension can be infinite, with absolutely no problem from a three-dimensional vantage point. Because the graviton makes only infrequent excursions into the bulk, a second brane or a curled-up dimension isn't necessary to get a theory that describes our three-dimensional world, as had previously been thought. We might live on the Planck brane and address the hierarchy problem in some other manner—or we might live on a second brane out in the bulk, but this brane would not be the boundary of the now infinite space. It doesn't matter that the graviton occasionally leaks away from the Planck brane; it's so highly localized there that the Planck brane essentially mimics a world of three dimensions, as though an extra dimension didn't exist at all. A four-spatial-dimensions world, say, would look almost identical to one with three spatial dimensions. Thus all the evidence we have for three spatial dimensions could equally well be evidence for a theory in which there are four spatial dimensions of infinite extent.

It's an exciting but frustrating game. We used to think the easiest thing to rule out would be large extra dimensions, because large extra dimensions would be associated with low energies, which are more readily accessible. Now, however, because of the curvature of space, there is a theory permitting an infinite fourth dimension of space in a configuration that so closely mimics three dimensions that the two worlds are virtually indistinguishable.

If there are differences, they will be subtle. It might turn out that black holes in the two worlds would behave differently. Energy can leak off the brane, so when a black hole decays it might spit out particles into the extra dimension and thus decay much more quickly. Physicists are now doing some interesting work on what black holes would look like if this extra-dimensional theory with the highly concentrated graviton on the brane is true;

however, initial inquiries suggest that black holes, like everything else, would look too similar to distinguish the four- and three-dimensional theories. With extra dimensions, there are an enormous number of possibilities for the overall structure of space. There can be different numbers of dimensions and there might be arbitrary numbers of branes contained within. Branes don't even all have to be three-plus-one-dimensional; maybe there are other dimensions of branes in addition to those that look like ours and are parallel to ours. This presents an interesting question about the global structure of space, since how space evolves with time would be different in the context of the presence of many branes. It's possible that there are all sorts of forces and particles we don't know about that are concentrated on branes and can affect cosmology.

In the above example, physics everywhere—on the brane and in the bulk—looks three-dimensional. Even away from the Planck brane, physics appears to be three-dimensional, albeit with weaker gravitational coupling. Working with Andreas Karch (now at the University of Washington), I discovered an even more amazing possibility: Not only can there be an infinite extra dimension but physics in different locations can reflect different dimensionality. Gravity is localized near us in such a way that it's only the region near us that looks three-dimensional; regions far away reflect a higher-dimensional space. It may be that we see three spatial dimensions not because there really are only three spatial dimensions but because we're stuck to this brane and gravity is concentrated near it, while the surrounding space is oblivious to our lower-dimensional island. There are also some possibilities that matter can move in and out of this isolated four-dimensional region, seeming to appear and disappear as it enters and leaves our domain. These are very hard phenomena to detect in practice, but theoretically there are all sorts of interesting questions about how such a construct all fits together.

Whether or not these theories are right will not necessarily be answered experimentally but could be argued for theoretically, if one or more of them ties into a more fundamental theory. We've used the basic elements found in string theory—namely, the existence of branes and extra dimensions—but we would really like to know if there is a true brane construction. Could you take the very specific branes given by string theory and produce a universe with a brane that localizes gravity? Whether you can actually derive this from string theory or some more fundamental theory is important. The fact that we haven't done it yet isn't evidence that it's not true, and Andreas and I have made good headway into realizing our scenario in string theory. But it can be very, very hard to solve these complicated geometrical set-ups. In general, the problems that get solved, although they seem very complicated, are in many ways simple problems. There's much more work to be done; exciting discoveries await, and they will have implications for other fields.

In cosmology, for instance. Alan Guth's mechanism whereby exponential expansion smooths out the universe works very well, but another possibility has been suggested: a cyclic universe, Paul Steinhardt's idea, wherein a smaller amount of exponential expansion happens many times. Such a theory prompts you to ask questions. First of all, is it really consistent with what we see? The jury's out on that. Does it really have a new mechanism in it? In some sense, the cyclic idea still uses inflation to smooth out the universe. Sometimes it's almost too easy to come up with theories. What grounds your theories? What ties them down? What restricts you from just doing anything? Is there really a new idea there? Do we really have a new mechanism at work? Does it connect to some other, more fundamental theoretical idea? Does it help make that work? Recently I have been exploring the implications of extra dimensions for cosmology. It seems that inflation with extra dimensions works even better than without! What's so nice about this theory is that one can reliably calculate the effect of the extra dimension; no ad-hoc assumptions are required. Furthermore, the theory has definite implications for cosmology experiments. All along, I've been emphasizing what we actually see. It's my hope that time and experiments will distinguish among the possibilities.

Lisa Randall

6
The Cyclic Universe

Neil Turok

Theoretical physicist, director, and Mike and Ophelia Lazardis
Neils Bohr Chair in Theoretical Physics, Perimeter Institute
for Theoretical Physics, Waterloo, Ontario; coauthor (with
Paul Steinhardt), *Endless Universe: Beyond the Big Bang*

For the last ten years, I have mainly been working on the question of how the universe began—or didn't begin. What happened at the Big Bang? To me, this seems like one of the most
fundamental questions in science, because everything we know
of emerged from the Big Bang. Whether it's particles or planets
or stars or, ultimately, even life itself.

In recent years, the search for the fundamental laws of nature
has forced us to think about the Big Bang much more deeply.
According to our best theories, string theory and M-theory, all
of the details of the laws of physics are determined by the structure of the universe—specifically, by the arrangement of tiny,
curled-up extra dimensions of space. This is a very beautiful picture: Particle physics itself is now just another aspect of cosmology. But if you want to understand why the extra dimensions
are arranged as they are, you have to understand the Big Bang,
because that's where everything came from.

Somehow, until quite recently, fundamental physics had gotten along without really tackling that problem. Even back in
the 1920s, Einstein, Friedmann, and Lemaître, the founders of
modern cosmology, realized there was a singularity at the Big

Bang. That somehow, when you trace the universe back, everything went wrong about 14 billion years ago. By "go wrong," I mean all the laws of physics break down: They give infinities and meaningless results. Einstein himself didn't interpret this as the beginning of time; he just said, "Well, my theory fails." Most theories fail in some regime, and then you need a better theory. Isaac Newton's theory fails when particles go very fast; it fails to describe that. You need relativity. Likewise, Einstein said we need a better theory of gravity than mine.

But in the 1960s, when the observational evidence for the Big Bang became very strong, physicists somehow leapt to the conclusion that it must have been the beginning of time. I'm not sure why they did so, but perhaps it was due to Fred Hoyle, the main proponent of the rival Steady State theory, who seems to have successfully ridiculed the Big Bang theory by saying it didn't make sense because it implied a beginning of time and that sounded nonsensical.

Then the Big Bang was confirmed by observation. And I think everyone just bought Hoyle's argument and said, "Oh, well, the Big Bang is true, OK, so time must have begun." So we slipped into this way of thinking—that somehow time "began," and that the process, or event, whereby it began is not describable by physics. That's very sad. Everything we see around us rests completely on that event, and yet that's the event we can't describe. That's basically where things stood in cosmology, and people just worried about other questions for the next twenty years.

And then in the 1980s there was a merging of particle physics and cosmology, when the theory of inflation was invented. Inflationary theory also didn't deal with the beginning of the universe, but it took us back further toward it. People said, "Let's just assume the universe began somehow, but we're going to assume that when it began it was full of a weird sort of energy called inflationary energy. This energy is repulsive—its gravitational field is not attractive, like ordinary matter—and the main property of that energy is that it causes the universe to expand, hugely fast. Literally like dynamite, it blows up the universe."

This inflationary theory became very popular. It made some predictions about the universe, and recent observations are very much in line with them. The type of predictions it made are rather simple and qualitative descriptions of certain features of the universe: It's very smooth and flat on large scales, and it has some density variations, of a very simple character. Inflationary theory predicts that the density variations are like random noise—something like the ripples on the surface of the sea— and fractional variation in the density is roughly the same on all length scales. And these predictions of inflation have been broadly confirmed by observation. So people have become very attracted to inflation, and many people think it's correct. But inflationary theory never really dealt with the beginning of the universe. We just had to *assume* that the universe started out full of inflationary energy. That was never explained.

My own work in this subject started about ten years ago, when I moved to Cambridge from Princeton. There I met Stephen Hawking, who, with James Hartle, developed a theory about how the universe can begin. So I started to work with Stephen, to do calculations to figure out what this theory actually predicted. Unfortunately, we quickly reached the conclusion that the theory predicted an empty universe. Indeed, this is perhaps not so surprising: If you start with nothing, it makes more sense that you'd get an empty universe rather than a full one. I'm being facetious, of course, but when you go through the detailed math, Hawking's theory seems to predict an empty universe, not a full one.

So we tried to think of various ways in which this problem might be cured, but everything we did to improve that result—to make the prediction more realistic—spoils the beauty of the theory. Theoretical physics is really a wonderful subject, because it's a discipline where crime does not pay in the long run. You can fake it for a while, you can introduce fixes and little gadgets which make your theory work, but in the long run, if it's no good, it'll fall apart. We know enough about the universe and the laws of nature, and how it all fits together, that it's extremely difficult to make a fully consistent theory. And when you start to cheat, you start to violate special symmetries which are, in fact, the key to the consistency of the whole structure. If those symmetries fall apart, then the whole theory falls apart. Hawking's theory is still an ongoing subject of research, and people are still working on it and trying to fix it, but I decided after four or five years that the approach wasn't working. It's very, very hard to make a universe begin and be full of inflationary energy. We needed to try something radically different.

So, along with Paul Steinhardt, I decided to organize a workshop at the Isaac Newton Institute in Cambridge, devoted to fundamental challenges in cosmology. And this was the big one: How to sensibly explain the Big Bang. We decided to bring together the most creative theorists in string theory, M-theory, and cosmology to brainstorm and see if there could be a different approach. The workshop was very stimulating, and our own work emerged from it.

String theory and M-theory are precisely the kinds of theories which Einstein himself had been looking for. His theory of gravity is a wonderful theory and still the most beautiful and successful theory we have, but it doesn't seem to link properly with quantum mechanics, which we know is a crucial ingredient for all the other laws of physics. If you try to quantize gravity naïvely, you get infinities which cannot be removed without spoiling all of the theory's predictive power. String theory succeeds in linking gravity and quantum mechanics within what seems to be a consistent mathematical framework. Unfortunately, thus far, the only cases where we can really calculate well in string theory are not physically realistic: For example, one can do precise calculations in static, empty space with some gravitational waves. Nevertheless, because of its tight and consistent mathematical structure, many people feel string theory is probably on the right track.

String theory introduces some weird new concepts. One is that every particle we see is actually a little piece of string. Another is that there are objects called branes, short for "membranes," which are basically higher dimensional versions of string. At the time of our workshop, a new idea had just emerged: the idea that the three dimensions of space we experience could in fact be the dimensions along one of these branes. The brane we live on could be a sort of sheet-like object floating around in a higher dimension of space. This underlies a model of the universe that fits particle physics very well and consists of two parallel branes separated by a very, very tiny gap. Many people were talking about this model in our workshop, including Burt Ovrut, and Paul and I asked the question of what happens if these two branes collide. Until then, people had generally only considered a static set-up. They described the branes sitting there, with particles on them, and they found that this set-up fit a lot of the data we have about particles and forces very well. But they hadn't considered the possibility that branes could move, even though that's perfectly allowed by the theory. And if the branes can move, they can collide. Our initial thought was that if they collide, that might have been the Big Bang. The collision would be a very violent process, in which the clash of the two branes would generate lots of heat and radiation and particles . . . just like a Big Bang.

Burt, Paul, and I began to study this process of the collision of the branes carefully. We realized that if it worked, this idea would imply that the Big Bang was not the beginning of time but, rather, a perfectly describable physical event. We also realized that this might have many implications if it were true. For example, not only could we explain the Bang but we could also explain the production of radiation which fills the universe, because there was a previous existing universe within which these two branes were moving. And what explained that, you might ask? That's where the cyclic model came in. The cyclic model emerged from the idea that each Bang was followed by another, and that this could go on for eternity. The whole universe might have existed forever, and there would have been a series of these Bangs, stretching back into the infinite past and into the infinite future.

For the last five years, we've worked on refining this model. The first thing we had to do was to match the model to observation, to see if it could reproduce some of the inflationary model's successes. Much to our surprise, we found that it could, and in some cases in a more economical way than inflation. If the two branes attract one another, then as they pull toward one another they acquire ripples, like the ripples on the sea I mentioned before. Those ripples turn into density variations as the branes collide and release matter and radiation, and these density variations later lead to the formation of galaxies in the universe.

We found that with some simple assumptions, our model could explain the observations to just the same accuracy as the inflationary model. That's instructive, because it says there are

Neil Turok

these two very different mechanisms which achieve the same end. Both models explain rather broad, simple features of the universe: that it's nearly uniform on large scales. That it's flat, like Euclidean space, and that it has these simple density variations, with nearly the same strength on every length scale. These features are explained either by the brane collision model or by the inflation model. And there might even be another, better model which no one has yet thought of. In any case, it's a healthy situation for science to have rival theories, which are as different as possible. This helps us to clearly identify which critical tests—be they observational or mathematical/logical—will be the key to distinguishing the theories and proving some of them wrong. Competition between models is good: It helps us see what the strengths and weaknesses of our theories are.

In this case, a key battleground between the more established inflationary model and our new cyclic model is theoretical: each model has flaws and puzzles. What happened before inflation? Does most of the universe inflate, or only some of it? Or, for the cyclic model, can we calculate all the details of the brane collision and turn the rough arguments into precise mathematics? It is our job as theorists to push those problems to the limit to see whether they can be cured or whether they will instead prove fatal for the models.

Equally, if not more important, is the attempt to test the models observationally, because science is nothing without observational test. Even though the cyclic model and inflation have similar predictions, there's at least one way we know of telling them apart. If there was a period of inflation—a huge burst of expansion just after the beginning of the universe—it would have filled space with gravitational waves, and those gravitational waves should be measurable in the universe today. Several experiments are already searching for them, and next year the European Space Agency's *Planck* satellite will make the best attempt yet: It should be capable of detecting the gravitational waves predicted by the simplest inflation models. Our model with the colliding branes predicts that the *Planck* satellite and other similar experiments will detect nothing. So we can be proved wrong by experiment.

Something I'm especially excited about right now is that we've been working on the finer mathematical details of what happens at the Bang itself. We've made some very good progress in understanding the singularity, where, according to Einstein's theory, everything becomes infinite—where all of space shrinks to a point, so the density of radiation and matter go to infinity and Einstein's equations fall apart.

Our new work is based on a very beautiful discovery made in string theory about ten years ago, with a technical name. It's called the Anti–de Sitter Conformal Field Theory correspondence. I won't attempt to explain that, but basically it's a beautiful geometrical idea which says that if I've got a region of space and time, which might be very large, then in some situations I can imagine this universe surrounded by what we call a boundary—which is basically a box enclosing the region we're interested in. About ten years ago, it was shown that even though the interior of this container is described by gravity, with all of the difficulties that brings, like the formation of black holes and the various paradoxes they cause, all of that stuff going on inside the box can be described by a theory that lives on the walls of the box surrounding the interior. That's the correspondence. A gravitational theory corresponds to another theory which has no gravity, and which doesn't have any of those gravitational paradoxes. What we've been doing recently is using this framework to study what happens at a cosmic singularity that develops in time, within the container. We study the singularity indirectly by studying what happens on the surface of the box surrounding the universe. When we do this, we find that if the universe collapses to make a singularity, it can bounce, and the universe can come back out of the bounce. As it passes through the singularity, the universe becomes full of radiation—very much like what happens in the colliding brane model—and density variations are created.

This is new work, but once it's completed I think it will go a long way toward convincing people that the Big Bang, or events like it, are actually describable mathematically. The model we're studying is not physically realistic, because it's a universe with four large dimensions of space. It turns out that's the easiest case to do, for rather technical reasons. Of course, the real universe has only three large dimensions of space, but we're settling for a four-dimensional model for the moment because the math is easier. Qualitatively, what this study is revealing is that you can study singularities in gravity and make sense of them. I think that's very exciting and I think we're on a very interesting track. I hope we will really understand how singularities form in gravity, how the universe evolves through them, and how those singularities go away.

I suspect that will be the explanation of the Big Bang—that the Big Bang was the formation of a singularity in the universe. I think by understanding it we'll be better able to understand how the laws of physics we currently see were actually set in place: why there's electromagnetism, the strong force, the weak force, and so on. All these things are a consequence of the structure of the universe, on small scales, and that structure was set at the Big Bang. It's a very challenging field, but I'm very happy we're actually making progress.

The current problem that's dominating theoretical physics—wrongly, I believe, because I think people ought to be studying the singularity and the Big Bang, since that's clearly where everything came from, but most people are just avoiding that problem—is the fact that the laws of physics we see, according to string theory, are a result of the specific configuration of the extra dimensions of space. So you have three ordinary dimensions, that we're aware of, and then there are supposed to be six more dimensions in string theory, which are curled up in a tiny little ball. At every point in our world, there would be another six dimensions, but twisted up in a tiny little knot. And the problem is that there are a huge number of ways of twisting up these extra dimensions. Probably there are an infinite number of ways. Roughly speaking, you can wrap them up by wrapping branes and other objects around them, twisting them up like a handkerchief with lots of bits of string and elastic bands wound around.

This caused many people to pull their hair out. String theory was supposed to be a unique theory and to predict one set of laws of physics, but the theory allows for many different types of universes with the extra dimensions twisted up in different ways. Which one do we live in? What some people have been doing, because they assume the universe simply starts after the Bang at some time, is just throwing dice. They say, "OK, well it could be twisted up in this way, or that way, or the other way, and we have no way of judging which one is more likely than the other, so we'll assume it's random." As a result, they can't predict anything. Because they don't have a theory of the Big Bang, they don't have a theory of why those dimensions ended up the way they are. They call this the landscape; there's a landscape of possible universes, and they accept that they have no theory of why we should live at any particular place in the landscape. So what do they do?

"Well," they say, "maybe we need the anthropic principle." The anthropic principle says the universe is the way it is because if it were any different we wouldn't be here. The idea is that there's this big landscape with lots of universes in it, but the only one that can allow us to exist is the one with exactly the laws of physics that we see. It sounds like a flaky argument—and it is. It's a very flaky argument. Because it doesn't predict anything. It's a classic example of postdiction: It's just saying, "Oh, well, it has to be this way, because otherwise we wouldn't be here talking about it." There are many other logical flaws in the argument which I could point to, but the basic point is that this argument doesn't really get you anywhere. It's not predictive and it isn't testable. The anthropic principle, as it's currently being used, isn't really leading to any progress in the subject. Even worse than that, it is discouraging people from tackling the important questions, like the fact that string theory, as it is currently understood, is incomplete and needs to be extended to deal with the Big Bang. That's just such an obvious point, but at the moment surprisingly few people seem to appreciate it.

I'm not convinced the landscape is real. There are still some reasonable mathematical doubts, about whether all these twisted-up configurations are legitimate. It's not been proven. But if it's true, then how are you going to decide which one of those configurations is adopted by the universe? It seems to me that whatever you do, you have to deal with the Big Bang. You need a mathematical theory of how Big Bangs works, either one that describes how time began or one that describes how the universe passes through an event like the Big Bang and, as it passes through, there's going to be some dramatic effect on these twisted-up dimensions. To me, the most plausible resolution of a landscape problem would be that the dynamics of the universe will select a certain configuration as the most efficient one for passing through Big Bangs and allowing a universe that cycles for a very long time.

For example, just to give a trivial example: If you ask, "Why is the gas in this room smoothly distributed," we need a physical theory to explain it. It wouldn't be helpful to say: "Well, if it wasn't that way, there would be a big vacuum in part of the room and if I walked into it, I would die. If the distribution of gas wasn't completely uniform, we wouldn't last very long." That's the anthropic principle. But it's not the scientific explanation. The explanation is that molecules jangle around the room, and when you understand their dynamics you understand that it's vastly more probable for them to settle down in a configuration where they're distributed nearly uniformly. It's nothing to do with the existence of people.

In the same way, I think the best way to approach the cosmological puzzles is to begin by understanding how the Big Bang works. Then, as we study the dynamics of the Bang, we'll hope to discover that the dynamics lead to a universe something like ours. If you can't understand the dynamics, you really can't do much except give up and resort to the anthropic argument. It's an obvious point, but strangely enough it's a minority view. In our subject, the majority view at the moment is this rather bizarre landscape picture, where somebody, or some random process, and no one knows how it happens, chooses for us to be in one of these universes.

The idea behind the cyclic universe is that the world we experience, the three dimensions of space, is actually an extended object which you can picture as a membrane, as long as you re-member that it's three-dimensional and we just draw it as two-dimensional because that's easier to visualize. According to this picture, we live on one of these membranes, and this membrane is not alone, there's another partner membrane, separated from it by a very tiny gap. There are three dimensions of space within a membrane, and a fourth dimension separating the two mem-branes. It so happens that in this theory there are another six dimensions of space, also curled up in a tiny little ball, but let's forget about those for the moment.

So you have this set-up with these two parallel worlds, just literally geometrically parallel worlds, separated by a small gap. We did not dream up this picture. This picture emerges from the most sophisticated mathematical models we have of the funda-mental particles and forces. When we try to describe reality—quarks, electrons, photons, and all these things—we are led to this picture of the two parallel worlds separated by a gap, and our starting point was to assume that this picture is correct.

These membranes are sometimes called end-of-the-world branes. Basically because they're more like mirrors; they're re-flectors. There's nothing outside them. They're literally the end of the world. If you traveled across the gap between the two membranes, you would hit one of them and bounce back from it. There's nothing beyond it. So all you have are these two par-allel branes with the gap. But these two membranes can move. So imagine we start from today's universe. We're sitting here, today, and we're living on one of these membranes. There's this other membrane, very near to us. We can't see it because light only travels along our membrane, but the distance away from us is much tinier than the size of an atomic nucleus. It's hardly any distance from us at all. We also know that, in the universe today, there's something called dark energy. Dark energy is the energy of empty space. Within the cyclic theory, the energy associated with the force of attraction between these two membranes is responsible, in part, for the dark energy.

Imagine that you've got these two membranes, and they attract each other. When you pull them apart you have to put energy into the system. That's the dark energy. And the dark energy itself causes these two membranes to attract. Right now the universe is full of dark energy; we know that from observations. According to our model, the dark energy is actually not stable, and it won't last forever. If you think of a ball rolling up a hill, the stored energy grows as the ball gets higher. Likewise, the dark energy grows as the gap between membranes widens. At some point, the ball turns around and falls back downhill. Likewise, after a period of dark-energy domination, the two branes start to move toward each other, and then they collide, and that's the Bang. It's the decay of the dark energy we see today which leads to the next Big Bang, in the cyclic model.

Dark energy was only observationally confirmed in 1999 and it was a huge surprise for the inflationary picture. There's no rhyme or reason for its existence in that picture: Dark energy plays no role in the early universe, according to inflationary theory. Whereas in the cyclic model, dark energy is vital, because it's the decay of dark energy that leads to the next Big Bang.

This picture of cyclic brane collisions resolves one of the longest-standing puzzles in cyclic models. The idea of a cyclic model isn't new: Friedmann and others pictured a cyclic model back in the 1930s. They envisaged a finite universe that collapsed and bounced over and over again. But Richard Tolman soon pointed out that, actually, it wouldn't remove the problem of having to have a beginning. The reason those cyclic models didn't work is that every bounce makes more radiation and that means the universe has more stuff in it. According to Einstein's equations, this makes the universe bigger after each bounce, so that every cycle lasts longer than the one before it. But, tracing back to the past, the duration of each bounce gets shorter and shorter and the duration of the cycles shrinks to zero, meaning that the universe still had to begin a finite time ago. An eternal cyclic model was impossible, in the old framework. What is new about our model is that by employing dark energy, and by having an infinite universe that dilutes away the radiation and matter after every bang, you actually can have an eternal cyclic universe, which could last forever.

7
Why Does the Universe Look the Way It Does?

Sean Carroll

Theoretical physicist, Caltech; author, *The Particle at the End of the Universe*

This seems, on the one hand, a very obvious question. On the other hand, it is an interestingly strange question, because we have no basis for comparison. The universe is not something that belongs to a set of many universes. We haven't seen different kinds of universes so we can say, "Oh, this is an unusual universe," or "This is a very typical universe." Nevertheless, we do have ideas about what we think the universe should look like if it were "natural," as we say in physics. Over and over again, it doesn't look natural. We think this is a clue to something going on that we don't understand.

One very classic example that people care a lot about these days is the acceleration of the universe and dark energy. In 1998, astronomers looked out at supernovae that were very distant objects in the universe, and they were trying to figure out how much stuff there was in the universe, because if you have more and more stuff—if you have more matter and energy—the universe would be expanding, but ever more slowly as the stuff pulled together. What they found by looking at these distant bright objects of type 1A supernovae was that not only is the universe expanding but it's accelerating. It's moving apart faster and faster. Our best explanation for this is something called dark energy—the idea that in every cubic centimeter of space, every little region of space, if you empty it out so there are no atoms, no dark matter, no radiation, no visible matter, there is still energy there. There is energy inherent in empty spaces. We can measure how much energy you need in empty space to fit this data, this fact that the universe is accelerating. This vacuum energy pushes on the universe. It provides an impulse. It keeps the universe accelerating. We get an answer, and the answer is 10^{-8} ergs per cubic centimeter, if that's very meaningful.

But then we can also estimate how big it should be. We can ask, "What should the vacuum energy have been?" We can do a back-of-the-envelope calculation, just using what we know about quantum field theory, the fact that there are virtual particles popping in and out of existence. We can say there should be a certain amount of vacuum energy. The answer is, there should be 10^{112} ergs per cubic centimeter. In other words, 10^{120} times as much as the theoretical prediction compared to the observational reality. That's an example where we say the universe isn't natural. There's a parameter of the universe, there's a fact about the universe in which we live—how much energy there is in empty space—which doesn't match what you would expect, what you would naïvely guess.

This is something a lot of attention has been paid to in the last ten years, or even before that—trying to understand the apparently finely tuned nature of the laws of physics. People talk about the anthropic principle and whether or not you could explain this by saying that if the vacuum energy were bigger we wouldn't be here to talk about it. Maybe there's a selection effect that says you can only live in a universe with finely tuned parameters like this. But there's another kind of fine tuning, another kind of unnaturalness, which is the state of the universe, the particular configuration we find the universe in—both now and at earlier times. That's where we get into entropy and the arrow of time.

This is actually the question I'm most interested in right now. It's a fact about the universe in which we observe that there are all sorts of configurations in which the particles in the universe could be. We have a pretty quantitative understanding of ways you could rearrange the ingredients of the universe to make it look different. According to what we were taught in the 19th century about statistical mechanics by Boltzmann and Maxwell and Gibbs and giants like that, what you would expect in a natural configuration is for something to be high entropy—for something to be very, very disordered. Entropy is telling us the number of ways you could rearrange the constituents of something so that it looks the same. In air filling the room, there are a lot of ways you could rearrange the air so that you wouldn't notice. If all the air in the room were squeezed into one tiny corner, there are only a few ways you could rearrange it. If air is squeezed into a corner, it's low entropy. If it fills the room, it's high entropy. It's very natural that physical systems go from low entropy, if they are low entropy, to being high entropy. There are just a lot more ways to be high entropy.

If you didn't know any better, if you asked what the universe should be like, what configuration it should be in, you would say it should be in a high-entropy configuration. There are a lot more ways to be high entropy; there are a lot more ways to be disorderly and chaotic than there are to be orderly and uniform and well arranged. However, the real world is quite orderly. The entropy is much, much lower than it could be. The reason for this is that the early universe, near the Big Bang, 14 billion years ago, had incredibly low entropy compared to what it could have been. This is an absolute mystery in cosmology. This is something that modern cosmologists don't know the answer to: why our observable universe started out in a state of such pristine regularity and order, such low entropy. We know that if it does, it makes sense. We can tell a story that starts in the low-entropy early universe, trace it through the present day and into the future. It's not going to go back to being low-entropy. It's going to be compliant entropy. It's going to stay there forever. Our best model of the universe right now is one that began 14 billion years ago in a state of low entropy but will go on forever into the future in a state of high entropy.

Why do we find ourselves so close to the aftermath of this very strange event, this Big Bang, that has such low entropy? The answer is, we just don't know. The anthropic principle is just not enough to explain this. We really need to think deeply about what could have happened both at the Big Bang and even before the Big Bang. My favorite guess at the answer is that the reason the universe started out at such a low entropy is the same reason that an egg starts out at low entropy. The classic example of entropy is that you can take an egg and make an omelet. You cannot take an omelet and turn it into an egg. That's because the entropy increases when you mix up the egg to make it into an omelet. Why did the egg start with such a low entropy in the first place? The answer is that it's not alone in the universe. The universe consists of more than just an egg. The egg came from a chicken. It was created by something that had a very low entropy that was part of a bigger system. The point is that our universe is part of a bigger system. Then you can start to try to understand why it had such a low entropy to begin with. I actually think that the fact that we can observe the early universe having such a low entropy is the best evidence we currently have that we live in a multiverse, that the universe we observe is not all there is, that we are actually embedded in some much larger structure.

We are in a very unusual situation in the history of science where physics has become slightly a victim of its own success. We have theories that fit the data, which is a terrible thing to have when you're a theoretical physicist. You want to be the one who invents those theories, but you don't want to live in a world where those theories have already been invented because then it becomes harder to improve upon them when they just fit the data. What you want are anomalies given to us by the data that we don't know how to explain.

Right now, we have two incredibly successful models—in fact three if you want to count gravity. We have for gravity Einstein's general theory of relativity, which we have had since 1915. It provides a wonderful explanation of how gravity works from the solar system to the very, very early universe—one second after the Big Bang. In particle physics, we have the standard model of particle physics, based on quantum field theory, and it predicts a certain set of particles. It was assembled over the course of the '60s and '70s, and then through the '80s and the '90s all we did was confirm that it was right. We got more and more evidence that it fit all of the data. The standard model is absolutely consistent with the observations that we had. Finally, in cosmology we have the standard Big Bang model—the idea that we start in a hot dense state near the Big Bang. We expand and cool over the course of 14 billion years. We have a theory for the initial conditions, where there were slight deviations in density from place to place and these slight deviations grow into galaxy and stars and clusters of galaxies.

The three ingredients—the standard model of particle physics, general relatively for gravity, and the standard model of Big

Bang cosmology—together fit essentially all the data we have. It makes it very difficult to move beyond that, but it's crucial that we move beyond that, because these ideas are mutually inconsistent with each other. We know they can't be the final answer. We have these large outstanding questions. How do you reconcile quantum field theory—and quantum mechanics more generally, which is the basis of the standard model—with general relativity, which is the way we describe gravity? These two theories are just speaking completely different languages, and that makes it very difficult to know how to marry them together. In cosmology, we have the Big Bang, which is a source of complete mystery. How did the universe begin? Why were the initial conditions like they were? That's something we need to figure out. We also have hints of things that don't quite fit into the model. We have dark matter, which cannot be accommodated in the standard model of particle physics, and we have dark energy making the universe accelerate, which is not something that we can do. We can basically put a fudge factor into the equations that fit the data, but again we don't have an understanding of why it is like that, where that comes from.

What we want to do is move beyond these models that fit the data, and are phenomenological and basically about fitting the data, and move to a deeper understanding. What are the fundamental ingredients out of which gravity and particle physics arise? What are the things that could have happened at the Big Bang? There are a bunch of ideas out there on the market.

For fundamental physics, we have string theory as the dominant paradigm. We don't know that string theory is right. It could be wrong, but for many years now people have been working on string theory, suggesting that we replace the idea of tiny little particles making up the universe by tiny little loops of string. That single idea taken to its logical conclusion predicts a whole bunch of wonderful things, which unfortunately we can't observe. This is just a problem with our ability to do experiments compared to the regime in which string theory might become important. We can't make a string by itself. We can't observe the stringiness of ordinary particles, because the energies are just too high. In string theory, you predict that there should be extra dimensions of space. There should not only be the three dimensions of space that we know and love—up, down, forward, backward— but there should be extra dimensions, and those dimensions are somehow invisible. They could all be curled up in really tiny balls and we just can't see them. In fact, we will never see them plausibly, depending on how small they are. Or some of them could be big, and we are stuck on some subset. We can't get to the extra dimensions, and this is the idea that we live on a brane. One of the questions that string theory puts front and center is, if the theory itself—string theory—likes to predict that there are extra dimensions, then where are they? Not only where are they, but why are they not visible? What happened in the universe to make these dimensions invisible? There are a bunch of ideas.

I recently wrote a paper with Lisa Randall and Matt Johnson about how we could have started in a universe that had more dimensions and then undergone a transition to a big space where some of the dimensions were curled up. This is a provocative idea that also feeds into cosmology. The point is that you can't just sit down and try to reconcile gravity—the laws of general relativity—with quantum mechanics without also talking about cosmology and why the universe started in the state that it was in. The way we have to go is to look at what happened before the Big Bang. Right now, the best model we have for what happened at what we now call the Big Bang, which is the favorite one among cosmologists, is inflation. The idea is that there was a temporary period of superfast acceleration that took a tiny little patch of the universe and smoothed it out, filled it with energy, and then that energy heated up into ordinary particles and dark matter, and that's what we see as the Big Bang today.

But inflation has a lot of questions that it doesn't answer. The most obvious question is, Why did inflation ever start? You say, "Well, there's a tiny little patch. It was dominated by some form of energy." How unlikely can that be? Roger Penrose and other people have emphasized that it's really, really unlikely that can be. Inflation doesn't provide a natural explanation for why the early universe looks like it does unless you can give me an answer for why inflation ever started in the first place. That's not a question we know the answer to right now. That's why we need to go back before inflation, into before the Big Bang, into a different part of the universe, to understand why inflation happened versus something else. There you get into branes and the cyclic universe.

I really don't like any of the models that are on the market right now. We really need to think harder about what the universe should look like. If we didn't have some prejudice for what the universe did look like from doing experiments, we should try to understand what we would expect just from first principles as to what the universe should look like, and then see how that comes close to, or is far away from, looking like the actual universe. It's only when we take seriously what our theories would like the universe to look like, and then try to match them with the universe that we see, that we can take advantage of these clues the experiments are giving us to try to reconcile the ideas of quantum mechanics, gravity, string theory, and cosmology.

One of the interesting things about the string theory situation, where we're victims of our own success, where we have models that fit the data very well but we're trying to move beyond them, is that the criteria for success has changed a little bit. It's not that one theory or another makes a prediction that you can go out and test tomorrow. We all want to test our ideas eventually, but it becomes a more long-term goal when it's hard to find data that doesn't already agree with the existing theories. We know that the existing theories aren't right and we need to move beyond them.

Quantum mechanics and general relativity are incompatible, but nature is not incompatible with itself. Nature figures out some way to reconcile these ideas. String theory is the obvious case of somewhere where it has been heavily investigated, starting in the '60s and '70s and taking off to become very popular in the '80s. Here we are in almost 2010 and it's still going strong without having made any connection to experiments. You might want to say at some point, "Show me the money." What have you actually learned from doing this? String theorists have learned a tremendous amount about string theory, but the question remains, Have we learned anything about nature? That's still an open question.

One of the reasons string theory is so popular among people who have thought about it carefully is that it really does lead to new things. It really is fruitful. It's not that you have to make some guess like, "Oh, maybe spacetime is discrete," or maybe the universe is made of little molecules, or something like that, and then you say, "OK, what do you get from that?" By making *this* guess, that instead of particles there are little strings, you're led to thinking, "If I put that into the framework of quantum mechanics I get ten dimensions." Then, "Oh, it also needs to be supersymmetric. There are different kinds of particles that we actually observe in nature, and if we try to compactify those extra dimensions and hide them, we begin to get things that look like the standard model. We're learning things that make us think we're on the right track."

In the 1990s there was a second superstring revolution that really convinced a lot of the skeptics that we *were* on the right track. There are still plenty of other skeptics who remain unconvinced. One of the things we learned is that different versions of string theory all come from the same underlying theory. Instead of there being many, many different versions of string theory, there's probably only one correct underlying theory that shows up in different ways. What you might have thought of as different versions of the universe, different versions of the laws of physics, are really more like different phases of matter. For water, we have liquid water, we have ice, we have water vapor. Depending on the conditions that the water is in, it will manifest itself in different ways and it will have different densities, different speed of sound, things like that. String theory says that's what spacetime can be like. Spacetime can find itself in different phases, like liquid water or frozen ice. In those different phases, the local laws of physics—the behavior around you—can look completely different. It can look dramatically different.

The most famous example was discovered by Juan Maldacena, a young string theorist who showed that you could have a theory in one version of which spacetime looked like gravity in five dimensions and in another version of which it looked like a four-dimensional theory without any gravity. There are different numbers of dimensions of space. In one version of the theory, there is gravity, and in another there is no gravity, but they're really the same theory under it all. To say that string theory [is a theory] of gravity is already not quite the whole story. It's a theory of some versions look like gravity, some versions you don't have any gravity.

The reason why that's so crucial is that there are a lot of philosophical problems that arise when you try to quantize gravity that don't arise when you try to talk about ordinary theories of particle physics without gravity. For example, the nature of time. Does the universe have a beginning? Do space and time emerge, or are they there from the start? These are very good questions, to which, *a priori*, we don't know the answer, but string theory has now given us a concrete, explicit playground, a toy example, where in principle all the answers are derivable. In practice, it might require a lot of effort to get there, but you can translate any question you have into a question in ordinary field theory without gravity. There is no beginning to time. Time and space are there, just as they are in ordinary particle physics.

We have learned a lot from string theory about what quantum gravity can be like. Whether or not it actually is, whether quantum gravity shows up in the real world, is still a little bit up for grabs. One of the problems is that it's easy to say you have different phases and that's interesting. The problem is that there are far too many phases. It's not like you have ten or twelve different possibilities and you need to match the right one onto the universe. It's like we have 10^{1000} different possibilities, or even maybe an infinite number of possibilities. Then you say, "Well, anything goes." You run into problems with falsifiability. How do you show that a theory is not right if you can get anything from it? My answer to that is, we just don't know yet. But that doesn't imply that we will never know.

The other thing is that we predict in string theory that there is a multiverse—that not only *can* you have different conditions in different places in the universe, but you *will*. If you combine ideas from string theory with ideas from inflation, you imagine that this universe we observe ourselves to be in is only a tiny lit-

Sean Carroll

tle part of a much, much larger structure where things are very different. People say, "Can you even talk about that and still call yourself a scientist? You talk about all this stuff we can never observe." The thing to keep in mind is that the multiverse is not a theory. The multiverse is a prediction of a theory. This theory that involves both string theory and inflation predicts that there should be regions outside what we can observe where conditions are very different. That's a crucially important difference, because we can imagine testing the theory in other ways even if we can't directly test the idea of the multiverse. The idea of the multiverse might change our expectations for why a certain thing we observe within our universe is a problem or not. It might say this issue about the small vacuum energy that we have isn't a problem, because we're just in one region of the universe that's not representative for one reason or another.

Basically, the short version of this long story is that we're on a long-term project here. We have very good ideas within string theory for reconciling quantum mechanics and gravity. We don't know if it's the right idea, but we're making progress. The fact that we don't yet know the answer—we can't yet make a firm falsifiable prediction for the Large Hadron Collider or for gravitational-wave observatories or for cosmology—is not in any way evidence that string theory is not on the right track. We have to both push forward with the experiments, get our hands dirty, learn more about cosmology, dark matter, and dark energy, and also push forward with the theories. Develop them to a point where we really can match them up to some experiment we haven't yet done.

We need some great ideas to be pushing forward from the condition we're in right now into the future, and my personal expertise is on the theoretical side of things. It is very often the case that the actual progress comes from the experimental side of things—and not just the experimental side of things but from experiments you hadn't anticipated were going to surprise you. There are these wonderful experiments people are doing—not only the big experiments with LIGO, detecting gravitational waves, and the LHC looking for new particles, but also smaller, table-top experiments, looking for small deviations from Newton's law of gravity, looking for new forces of nature that are very weak, or new particles that are very hard to detect. I need to say that it would be very likely that one of these experiments in some unanticipated way jolts us out of our dogmatic slumber and give us some new ideas.

I have an opinion, which is slightly heterodox, about the standard ideas in cosmology. The inflationary universe scenario that Alan Guth really pioneered, people like Andre Linde and Paul Steinhardt really pushed very hard—this is a wonderful idea, which I suspect is right. I suspect that some part of the history of the universe is correctly explained by the idea of inflation, the idea that we start in this little tiny region that expanded and accelerated at this superfast rate. However, I think that the way most people, including the people who invented the idea, think about inflation is wrong. They're too sanguine about the idea that inflation gets rid of all the problems that the early universe might have had. There's this feeling that inflation is like confession— that it wipes away all prior sins. I don't think that's right. We haven't explained what needs to be explained until we take seriously the question of why inflation ever started in the first place. It's actually a mistake, and something wrong on the part of many of the people who buy into inflation, that inflation doesn't need to answer that question because once it starts it answers all the questions you have.

When I was in graduate school happily reading all these different papers and learning different things, some of the papers I read were by Roger Penrose, who was a skeptic about the prevailing conventional wisdom concerning the inflationary universe scenario. Penrose kept saying over and over again in very clear terms that inflation doesn't answer the question we want answered because it doesn't explain why the early universe had a low entropy. It says why the universe evolved in the way it did by positing that the universe started in an even lower entropy state than was conventionally assumed. It's true that if you make that assumption, everything else follows, but there's no reason, Penrose said, to make that assumption. I read those papers and I knew that there was something smart being said there, but I thought Penrose had missed the point and so I basically dismissed him.

Then I read papers by Huw Price, who is a philosopher in Australia and who made basically the same point. He said that cosmologists are completely fooling themselves about the entropy of the universe. They're letting their models assume that the early universe had a low entropy, the late universe has a very high entropy. But there's no such asymmetry built into the laws of physics. The laws of physics at a deep level treat the past and the future the same. But the universe doesn't treat the past and the future the same. One way of thinking about it is, if you were out in space floating around, there would be no preferred notion of up or down, left or right. There is no preferred direction in space. Here on Earth, there's a preferred notion of up or down because there's the Earth beneath us. There is this dramatic physical object that creates a directionality to space, up versus down. Likewise, if you were in a completely empty universe, there would be no notion of past and future. There would be no difference between one direction of time or the other.

The reason we find a direction in time here in this room, or in the kitchen when you scramble an egg or mix milk into coffee, is not because we live in the physical vicinity of some important object but because we live in the aftermath of some influential event, and that event is the Big Bang. The Big Bang set all of the clocks in the world. When we get down to how we evolve, why we are born and then die, and never in the opposite order, why we remember what happened yesterday and we don't remember what's going to happen tomorrow, all these manifestations of the difference between the past and the future are coming from the same source. That source is the low entropy of the Big Bang.

This is something that was touched on way back in the 19th century, when the giants of thermodynamics like Boltzmann and Maxwell were trying to figure out how entropy works and how thermodynamics works. Boltzmann came up with a great definition of entropy, and he was able to show that if the entropy is low, it will go up. That's good because that's the second law of thermodynamics. But he was stuck on this question of why was the entropy low to begin with. He came up with all these ideas which are very reminiscent of the same kinds of ideas that cosmologists are talking about today. Boltzmann invented the idea of a multiverse, the anthropic principle, where things were different in some regions of the universe than in others and we lived in an unrepresentative part of it. But he never really quite settled on what he thought was the right answer, which makes perfect sense, because still today we don't know what the right answer is. We know very well how to explain that I remember yesterday and not tomorrow, but only if we assume that we start the universe in a low-entropy state.

I like to say that observational cosmology is the cheapest possible science to go into. Every time you put milk in your coffee and watch it mix and realize that you can't unmix that milk from your coffee, you're learning something profound about the Big Bang—about conditions in the very, very early universe. This is just a giant clue the real universe has given to us to how the fundamental laws of physics work. We don't yet know how to put that clue to work. We don't know the answer to the whodunnit, who is the guilty party, why the universe is like that. But taking this question seriously is a huge step forward in trying to understand how the universe we see around us directly fits into a much bigger picture.

8
In the Matrix

Martin Rees

Former president, the Royal Society; emeritus professor of cosmology and astrophysics, University of Cambridge; master, Trinity College; author, *From Here to Infinity*

This is a really good time to be a cosmologist, because in the last few years some of the questions we've been addressing for decades have come into focus. For instance, we can now say what the main ingredients of the universe are: It's made of 4-percent atoms, about 25-percent dark matter, and 71-percent mysterious dark energy latent in empty space. That's settled a question we've wondered about certainly the entire thirty-five years I've been doing cosmology.

We also know the shape of space. The universe is "flat"—in the technical sense that the angles of even very large triangles add up to 180 degrees. This is an important result that we couldn't have stated with confidence two years ago. So a certain phase in cosmology is now over.

But as in all of science, when you make an advance, you bring a new set of questions into focus. And there are really two quite separate sets of questions that we are now focusing on. One set of questions addresses the more environmental side of the subject; we're trying to understand how, from an initial Big Bang nearly 14 billion years ago, the universe has transformed itself into the immensely complex cosmos we see around us, of stars and galaxies, et cetera; how, around some of those, stars and planets arose;

and how, on at least one planet, around at least one star, a biological process got going and led to atoms assembling into creatures like ourselves, able to wonder about it all. That's an unending quest—to understand how the simplicity led to complexity. To answer it requires ever more computer modeling, and data in all wavebands from ever more sensitive telescopes.

Another set of questions that come into focus are the following:

• Why is the universe expanding the way it is?
• Why does it have the rather arbitrary mix of ingredients?
• Why is it governed by the particular set of laws which seem to prevail in it, and which physicists study?

These are issues where we can now offer a rather surprising new perspective. The traditional idea has been that the laws of nature are somehow unique; they're given, and are "there" in a Platonic sense independent of the universe which somehow originates and follows those laws.

I've been puzzled for a long time about why the laws of nature are set up in such a way that they allow complexity. That's an enigma, because we can easily imagine laws of nature which weren't all that different from the ones we observe but which would have led to a rather boring universe—laws which led to a universe containing dark matter and no atoms; laws where you perhaps had hydrogen atoms but nothing more complicated, and therefore no chemistry; laws where there was no gravity, or a universe where gravity was so strong that it crushed everything; or whose lifetime was so short that there was no time for evolution.

It always seemed to me a mystery why the universe was, as it were, "biophilic"—why it had laws that allowed this amount of complexity. To give an analogy from mathematics, think of the Mandelbrot Set; there's a fairly simple formula, a simple recipe that you can write down, which describes this amazingly complicated pattern, with layer upon layer of structure. Now, you could also write down other rather similar-looking recipes, similar algorithms, which describe a rather boring pattern. What has always seemed to me a mystery is why the recipe, or code, that determined our universe had these rich consequences, just as the algorithms of the Mandelbrot Set, rather than describing something rather boring in which nothing as complicated as us could exist.

For about twenty years I've suspected that the answer to this question is that perhaps our universe isn't unique. Perhaps, even, the laws are not unique. Perhaps there were many Big Bangs, which expanded in different ways, governed by different laws, and we are just in the one that has the right conditions. This thought in some respect parallels the way our concept of planets and planetary systems has changed.

People used to wonder, Why is the Earth in this rather special orbit around this rather special star, which allows water to exist, or allows life to evolve? It looks somehow fine-tuned. We now perceive nothing remarkable in this, because we know that there are millions of stars with retinues of planets around them: Among that huge number there are bound to be some that have the conditions right for life. We just happen to live on one of that small subset. So there's no mystery about the fine-tuned nature of the Earth's orbit; it's just that life evolved on one of millions of planets where things were right.

It now seems an attractive idea that our Big Bang is just one of many. Just as our Earth is a planet that happens to have the right conditions for life among the many, many planets that exist, so

Martin Rees

our universe, and our Big Bang, is the one out of many which happens to allow life to emerge, to allow complexity. This was originally just a conjecture, motivated by a wish to explain the apparent fine-tuning in our universe—and incidentally a way to undercut the so-called theological design argument, which said that there was something special about these laws.

But what's happened in the last few years, and particularly in the last year, is that the basis for this so-called multiverse idea has strengthened, and, moreover, the scale we envisage for the multiverse has got even vaster than we had in mind a few years ago. There's a firmer basis for the "multiverse" concept, because recent work on the best theory we have for the fundamental laws of nature—namely, superstring theory—suggests that there should indeed be many possible forms for a universe and many possible laws of nature.

At first it was thought that there might be just one unique solution to the equations, just one possible three-dimensional universe with one possible "vacuum state" and one set of laws. But it seems now, according to the experts, that there could be a huge number. In fact, Lenny Susskind claims that there could be more possible types of universe than there are atoms in our universe—a quite colossal variety. The system of universes could be even more intricate and complex than the biosphere of our planet. This really is a mind-blowing concept, especially when we bear in mind that each of those universes could themselves be infinite.

At first sight, you might get worried about an infinity of things in themselves infinite, but to deal with this you have to draw on a body of mathematics called transfinite number theory, which goes back to Cantor in the 19th century. Just as many kinds of pure mathematics have already been taken over by physicists, this rather arcane subject of transfinite numbers is now becoming relevant, because we've got to think of infinities of infinity. Indeed, there's perhaps even a higher hierarchy of infinities: In addition to our universe being infinite, and there being an infinite number of possible laws of nature, we may want to incorporate the so-called many-worlds theory of quantum mechanics. Each "classical" universe is then replaced by an infinite number of superimposed universes, so that when there's a quantum choice to be made, the path forks into extra universes. This immensely complicated construct is the consequence of ideas that are still speculative but are firming up. One of the most exciting frontiers of 21st-century physics is to utilize the new mathematics and the new cosmology to come to terms with all this.

What we've traditionally called "our universe" is just a tiny part of something which is infinite, so allows for many replicas of us elsewhere, in our same spacetime domain but far beyond the horizon of our observations. But even that infinite universe is just one element of an ensemble that encompasses an infinity of quite different universes. So that's the pattern adumbrated by cosmology and some versions of string theory. What we have normally called the laws of nature are not universal laws—they're just parochial by-laws in our cosmic patch, no more than that, and a variety of quite different regimes prevail elsewhere in the ensemble.

One thing which struck me recently, and I found it a really disconcerting concept, was that once we accept all that, we get into a very deep set of questions about the nature of physical reality. That's because even in our universe, and certainly in some of the others, there'd be the potential for life to develop far beyond the level it's reached on Earth today. We are probably not the culmination of evolution on Earth; the time lying ahead for the

Earth is as long as the time that's elapsed to get from single-celled organisms to us, and so life could spread, in a post-human phase, far beyond the Earth. In other universes there may be an even richer potentiality for life and complexity.

Now, life and complexity means information-processing power; the most complex conceivable entities may not be organic life but some sort of hypercomputers. But once you accept that our universe, or even other universes, may allow the emergence within them of immense complexity, far beyond our human brains, far beyond the kind of computers we can conceive, perhaps almost at the level of the limits that Seth Lloyd discusses for computers—then you get a rather extraordinary conclusion. These super- or hypercomputers would have the capacity to simulate not just a simple part of reality but a large fraction of an entire universe.

And then of course the question arises: If these simulations exist in far larger numbers than the universe themselves, could we be in one of them? Could we ourselves not be part of what we think of as bedrock physical reality? Could we be ideas in the mind of some supreme being, as it were, who's running a simulation? Indeed, if the simulations outnumber the universes, as they would if one universe contained many computers making many simulations, then the likelihood is that we are "artificial life" in this sense. This concept opens up the possibility of a new kind of virtual time travel, because the advanced beings creating the simulation can, in effect, rerun the past. It's not a time-loop in a traditional sense; it's a reconstruction of the past, allowing advanced beings to explore their history.

All these multiverse ideas lead to a remarkable synthesis between cosmology and physics. . . . But they also lead to the extraordinary consequence that we may not be the deepest reality, we may be a simulation. The possibility that we are creations of some supreme or superbeing blurs the boundary between physics and idealist philosophy, between the natural and the supernatural, and between the relation of mind and multiverse and raises the possibility that we're in the matrix rather than the physics itself.

Once you accept the idea of the multiverse, and that some universes will have immense potentiality for complexity, it's a logical consequence that in some of those universes there will be the potential to simulate parts of themselves, and you may get sort of infinite regress, so we don't know where reality stops and where the minds and ideas take over, and we don't know what our place is in this grand ensemble of universes and simulated universes.

Considerations of the multiverse change the way we think about ourselves and our place in the world. Traditional religion is far too blinkered to encompass the complexities of mind and cosmos. All we can expect is to have a very incomplete and metaphorical view of this deep reality. The gulf between mind and matter is something we don't understand at all, and [when] some minds can evolve to the stage that they can create other minds, there's real blurring between the natural and the supernatural.

My attitude towards religion is really twofold. First, as far as the practice of religion is concerned, I appreciate it and participate in it, but I'm skeptical about the value of interactive dialogue. There's no conflict between religion and science (except, of course, with naïve creationism and suchlike), but I doubt—unlike some members of the Templeton Foundation—that theological insights can help me with my physics. I'm fascinated to talk to philosophers (and with some theologians) about their work, but I don't believe they can help me very much. So I favor peaceful coexistence rather than constructive dialogue between science and theology.

I am concerned about the threats and opportunities posed by 21st-century science and how to react to them. There are some intractable risks stemming from science, which we have to accept as the downside for our intellectual exhilaration and—even more—for its immense and ever more pervasive societal benefits. I believe there's an expectation of a 50-percent chance of a really severe setback to civilization by the end of the century. Some people have said that's unduly pessimistic. But I don't think it is. Even if you just consider the nuclear threat, I think that's a reasonable expectation.

If we look back over the cold war era, we know we escaped devastation, but at the time of the Cuban crisis, as recent reminiscences of its 40th anniversary have revealed, we were really on a hair trigger, and it was only through the responsibility and good sense of Kennedy and Khrushchev and their advisers that we avoided catastrophe. Ditto on one or two other occasions during the cold war. And that could indeed have been a catastrophe. The nuclear arsenals of the superpowers have the explosive equivalent of one of the U.S. Air Force's daisy cutter bombs for each inhabitant of the United States and Europe. Utter devastation would have resulted had this all gone off.

The threat obviously abated at the end of the cold war, but looking a century ahead, we can't expect the present political assignments to stay constant. In the last century, the Soviet Union appeared and disappeared, and there were two world wars. Within the next hundred years, since nuclear weapons can't be disinvented, there's quite a high expectation that there will be another standoff as fearsome as the cold war era, perhaps involving more participants than just two and therefore more unstable. Even if you consider the nuclear threat alone, then there is a severe chance, perhaps a 50-percent chance, of some catastrophic setback to civilization.

There are other novel threats as well. Not only will technical change be faster in this century than before, but it will take place in more dimensions. Up to now, one of the fixed features over all recorded history has been human nature and human physique; human beings themselves haven't changed, even though our environment and technology has. In this century, human beings are going to change, because of genetic engineering, because of targeted drugs, perhaps even because of implants in their brain to increase our mental capacity. Much that now seems science fiction might, a century ahead, become science fact. Fundamental changes like that—plus the runaway development of biotech, possibly nanotechnology, possibly computers reaching superhuman intelligence—open up exciting prospects, but also all kinds of potential scenarios for societal disruption, even for devastation.

We have to be very circumspect if we are to absorb these rapid developments without a severe disruption. In the near term, the most severe threat stems from developments in biotechnology and genetic engineering. An authoritative report just last year by the National Academy of Sciences emphasized that large numbers of people would acquire the capability to engineer modified viruses for which existing vaccines might be ineffective, and thereby trigger some sort of epidemic.

The scary realization is that to do this would not require a large terrorist group, certainly not rogue states, but just an individual with the same mindset as an arsonist. Such people might start hacking real viruses. If an epidemic ensued, it probably could be contained in this country, but as the SARS episode shows, infections rapidly spread around the world, and if any of these

Martin Rees

epidemics reach the megacities of the third world then they could really take off. I took a public bet—one of the "long bets" set up by *Wired* magazine—that within twenty years one instance of bio error or bio terror would lead to a million fatalities. That's not unduly pessimistic: It would require just one weirdo to release a virus that spread into the third world. That's a scarifying possibility, rendered possible by technology that we know is going to become available. And it's difficult to control, as the kind of equipment that's needed is small-scale. Also, it's the kind of technology which is utilized for all kinds of benign purposes anyway and is bound to exist, unless we entirely stop drug developments and other biotechnology.

I can foresee what might happen if there were one such event like this that happened in the United States. Supposing that there were a mysterious epidemic which maybe didn't kill a huge number of people but which could be traced to some engineered virus, maliciously or erroneously released. Everyone would realize that if this happened once, it could happen again, anytime, anywhere. And they'd realize also that the only way to stop a repeat would be to have intrusive surveillance to keep tabs on everyone who had that expertise. There would be real pressure to infringe on basic freedoms, and a strong anti-science backlash, if one event like this happened. That's the near-term danger which I worry about most.

Being a cosmologist doesn't make me worry less than anyone else about what happens tomorrow or next week or next year. But it does give me a different perspective, because cosmologists are aware of the long-term future. Most educated people are now aware that we as humans are the outcome of billions of years of evolution. Almost 4 billion years of Darwinian selection separate us from the very first microorganisms. But most people nonetheless, at least subconsciously, feel that we as humans are a kind of culmination—that evolution led to us and that's that.

But anyone who's studied astronomy knows that the sun is less than halfway through its life and the universe may have an infinite future. So the time lying ahead for evolution is at least as long as the time elapsed up to now. The post-human phase of evolution could be at least as long as what has led from single-celled organisms to humans, and of course you only have to read science fiction to realize the scenarios whereby life can evolve here on the Earth in more elaborate ways, and, more likely, can spread beyond the Earth; life from the Earth could even "green" the entire galaxy, if given enough time. And that time does exist.

As a cosmologist, I'm more aware of the immensely long-term potential that we'd be foreclosing if we screwed things up here on Earth this century. This perspective gives us an extra motive for cherishing this pale blue dot in the cosmos, because of the importance it might have for the long-range future of life even beyond the Earth.

I've been in cosmology for thirty-five years now, and what has been a tremendous boost to my morale is that the pace of discovery has not slowed up at all. The 1960s seemed an exciting time. That's when the first evidence for the Big Bang appeared. It's also when the first high-redshift quasars were discovered, along with the first evidence of black holes, neutron stars, et cetera. It was good to be a young cosmologist, because when everything's new, the experience of the old guys is at a discount and all had to start afresh.

But what's happened in the last three or four years is just as exciting as in any previous period that I can remember. In cosmology we have not only lots of fascinating ideas about how the universe began, ideas of how complexity developed, the possibility of extra dimensions playing a role, et cetera. But we also have new evidence which pins down some of the key numbers of the universe. We know we live in a flat universe where the atoms that make up us, the stars, the planets, and the galaxies constitute only 4 percent of the mass and energy. About 25 percent is in mysterious dark matter, which helps with the gravitational binding of galaxies. And the remainder, 71 percent, is even more mysterious: some kind of energy latent in empty space itself. To explain dark matter is a challenge to physicists; it's probably some kind of particle left over from the Big Bang.

To explain dark energy is even more daunting. Superstring theorists believe it is the biggest challenge to their theory, because it tells us that the empty space we live in, our "vacuum," is something which isn't just nondescript; it has a well-defined energy, a well-defined tension in it, which affects the overall cosmic dynamics, causing an acceleration of the Hubble expansion.

Another important advance has been in understanding the emergence of structure within the universe. To put this in context, let's imagine how the universe evolved. It started off as a very hot fireball. As it expanded, it cooled down; the radiation diluted, and its wavelengths stretched. After about half a million years, the universe literally entered a dark age, because instead of glowing bright and blue the primordial heat, the heat of the fireball, then shifted into the infrared, and the universe became literally dark. This dark age continued until the first stars formed and lit it up again.

One of my long-standing interests has been trying to understand when and how this happened. We've had some new clues recently from observations using giant ground-based telescopes, from the WMAP satellite, and also from being able to do computer simulations of how the first structures formed. We are trying to combine theories and observations to understand the formative stages of galaxies—how the first stars formed, in units much smaller than galaxies, then they assembled together to make galaxies, and how simple atoms of hydrogen and helium gradually get transmuted in early generations of stars into carbon, oxygen, silicon, and iron, the building blocks of planets and then of life.

We need to understand how long it took to get the first planets, how long it took to get the first potentiality of life, how long to get the first big galaxies from these small precursor substructures. More clues are coming from observations using the most powerful telescopes we now have—the biggest of all is the European Southern Observatory's Very Large Telescope in Chile, which is really four telescopes, each with 8-meter mirrors, that can be linked together.

These huge mirrors allow us to detect very faint and distant objects. The further out you look in space, the further back you look in time. The goal is to look back far enough to actually see galaxies in their formative stages—even perhaps to see pregalactic eras when the first stars were formed.

That's a possibility that arises from another of my main interests, which is the most powerful explosion in the universe— gamma-ray bursts. These represent the violent end of a particular kind of supernova explosion. They're so powerful that they could be readily detected even from the era when the very first stars were born. If some of these very first stars end their lives as gamma-ray bursts, then we can use them to probe the earliest phases of galaxy formation—how the dark age ended and how the structures gradually built up. Big telescopes will offer snapshots of what the universe was like at various stages in the past.

There have been these amazing developments of the structure of the universe. But if I was asked to think of what else has been exciting in astronomy recently, the undoubted other highlight has been the discovery of large numbers of planets around other stars. Only in 1995 did astronomers find the first evidence for a planet orbiting another star like the sun. Now there are more than a hundred, and there's every expectation that a large proportion of the stars you can see in the sky have retinues of planets orbiting them.

Ten or twenty years from now, looking up at the sky will be a more interesting experience, because the stars won't just be twinkling points of light but for each of them we'll be able to say something about the planets it has orbiting around it, their masses, their orbits, and perhaps the topographical features of the largest planets. That will make the night sky a lot more interesting and make everyone appreciate the universe as a much richer and more diverse place. Most of those planets will be inhospitable to life, but astronomers will have found planets which are rather like the Earth. And that will give a focus for addressing the questions of life in the universe. We will be able to analyze the light from these Earth-like planets, and test whether there is, for instance, ozone in their atmospheres, which would be a signal of biological processes. And that would give us a clue as to whether there might be life. This enterprise will be complemented by progress by biologists in understanding the origin of life on Earth—by experiments and perhaps also computer simulations. I'm very hopeful that twenty years from now we'll understand the origin of life; we'll have a feel for whether life is widespread in the universe; we might be able to point to particular other planets orbiting other stars that might have life on them.

There is then a rather separate question: Whether simple life is likely to evolve into anything we might recognize as intelligent or complex. That may be harder to decide. Some people say there are many hurdles to be surmounted in going from simple life to complex life and life on Earth is lucky to surmount those hurdles, but others suspect that life would somehow find its way to great complexity. Among my friends and colleagues, there's a disparity of belief.

As an astronomer, I'm often asked, "Isn't it a bit presumptuous to try to say anything with any level of confidence about these vast galaxies?" or the Big Bang, et cetera. My response is that what makes things hard to understand isn't how big they are but how complicated they are. There's a real sense in which galaxies and the Big Bang and stars are quite simple things. They don't have the same intricate layer upon layer of structure that an insect does, for instance. And so the task of understanding the complexities of life is in some respects more daunting than the challenge of understanding our Big Bang and of understanding the microworld of atoms, as challenging as they are, too.

It's rather interesting that the most complicated thing to develop in the universe—namely, human beings—are, in a well-defined sense, midway between atoms and stars. It would take about as many human bodies to make up the mass of the sun as there are atoms in each of us. That's a surprisingly precise statement: The geometric mean of the mass of a proton and the mass of the sun is about 55 kilograms—not far off the mass of an average person. It's surprising that this is such a close coincidence, but it's not surprising that the most complicated things are on this intermediate scale between cosmos and microworld. Anything complicated has to be made of huge numbers of atoms with many layers of structures; it's got to be very, very big compared to an atom. On the other hand, there's a limit, because any structure that gets too big is crushed by gravity. You couldn't have a creature a mile high on the Earth—even Galileo realized that. And something as big as a star or a planet is completely molded by gravity and no internal structures survive, so it's clear that complexity exists on this intermediate scale.

Looking forward to the next decade, I expect development of the fundamental understanding of the Big Bang; I expect development in understanding the emergence and structure of the universe, using computer simulation observation; and I expect at least the beginnings of an integration between computer simulations, observations, and biological thought in the quest to understand how planets formed and how they developed biospheres.

Cosmology has remained lively and the focus of my interest. It's not only developing fast, and of fundamental importance, but it also has a positive and nonthreatening public image. That makes it different from other high-profile sciences like genetics and nuclear science, about which there's public ambivalence. It's also one in which there is wide public interest. I'd derive less satisfaction from my research if I could talk to only a few fellow specialists about it. It's a bonus that there's a wide public which is interested in origins. Just as Darwinism has been since the 19th century, cosmology and fundamental physics are now part of public culture. Darwin tried to understand how life evolved on this Earth. I and other cosmologists try to set the entire Earth into cosmic context—to trace the origin of the atoms that make it up, right back to a simple beginning in the Big Bang. There's public interest in how things began: Was there a beginning, will the universe have an end, and so forth.

It's good for us as researchers to address a wider public. It makes us realize what the big questions are. What I mean by this is that in science the right methodology is often to focus on a piece of the problem which you think you can solve. It's only cranks who try to solve the big problems at one go. If you ask scientists what they're doing, they won't say "Trying to cure cancer" or "Trying to understand the universe." They'll point at something very specific. Progress is made by solving bite-sized problems one at a time. But the occupational risk for scientists is that even though that's the right methodology, they sometimes lose sight of the big picture. Members of a lay audience always ask the big questions, the important questions, and that helps us to remember that our piecemeal efforts are only worthwhile insofar as they're steps towards answering those big questions.

Martin Rees

9
Think About Nature

Lee Smolin

Theoretical physicist, Perimeter Institute for Theoretical Physics; author, *Time Reborn*

The main question I'm asking myself, the question that puts everything together, is how to do cosmology, how to make a theory of the universe as a whole system. This is said to be the golden age of cosmology, and it is, from an observational point of view. But from a theoretical point of view it's almost a disaster. It's crazy, the kind of ideas we find ourselves thinking about. And I find myself wanting to go back to basics, to basic ideas and basic principles, and understand how we describe the world in a physical theory.

What's the role of mathematics? Why does mathematics come into physics? What's the nature of time? These two things are very related, since mathematical description is supposed to be outside of time. And I've come to a long evolution since the late 1980s, to a position that is quite different from the ones I had originally, and quite surprising even to me. But let me get to it bit by bit. Let me build up the questions and the problems that arise.

One way to start is with what I call "physics in a box," or theories of small isolated systems. The way we've learned to do this is to make an accounting, or an itinerary—a listing of the possible states of a system. How can a possible system be? What are the possible configurations? What were the possible states? If it's a glass of Coca-Cola, what are the possible positions and states of all the atoms in the glass? Once we know that, we ask, "How do the states change?" And the metaphor here—which comes from atomism, which comes from Democritus and Lucretius—is that physics is nothing but atoms moving in a void, and the atoms never change. The atoms have properties, like mass and charge, that never change in time. The void—which is space—in the old days never changed in time. It was fixed, and the atoms moved according to laws, which were originally given by, or tried to be given by, Descartes and Galileo, given by Newton much more successfully. And up until the modern era, when we describe them in quantum mechanics, the laws also never changed. The laws let us predict where the positions of the atoms will be at a later time if we know the positions of all the atoms at a given moment. That's how we do physics and I call that the Newtonian paradigm, because it was invented by Newton.

And behind the Newtonian paradigm is the idea that the laws of nature are timeless. They act on the system, so to speak, from outside the system, and they evolve from the past to the present to the future. If you know the state at any time, you can predict the state at any other time. So this is the framework for doing physics and it's been very successful. And I'm not challenging its success within its proper domain: small parts of the universe.

The problem that I've identified—that I think is at the root of a lot of the spinning of our wheels and confusion of contemporary physics and cosmology—is that you can't just take this method of doing science and scale it up to the universe as a whole. When you do, you run into questions that you can't answer. You end up with fallacies; you end up saying silly things. One reason is that on a cosmological scale, the questions we want to understand are not just "What are the laws?" but "Why are these the laws rather than other laws? Where do the laws come from? What makes the laws what they are?" And if the laws are input to the method, the method will never explain the laws, because they're input.

Also, given the state of the universe, of the system, at one time, we use the laws to predict the state at a later time. But what was the cause of the state that we started with that initial time? Well, it was something in the past, so we have to evolve from [a state] further in the past. And what was the reason for *that* past state? Well, that was something further and further in the past. So we end up at the Big Bang. It turns out that any explanation for why we're sitting in this room—why is the Earth in orbit around the sun where it is now—any question of detail that we want to ask about the universe ends up being pushed back, using the laws, to the initial conditions of the Big Bang.

And then we end up wondering, Why were those initial conditions chosen? Why that particular set of initial conditions? Now we're using a different language. We're not talking about particles and Newton's laws, we're talking about quantum field theory. But the question is the same: What chose the initial conditions? And since the initial conditions are input to this method that Newton developed, it can't be explained within that method. So if we want to ask cosmological questions, if we want to really explain everything, we need to apply a different method. We need to have a different starting point. And the search for that different method has been the central point in my thinking since the early '90s.

Now, some of this is not new. The American philosopher Charles Sanders Peirce identified this issue that I've just mentioned in the late 19th century. However, his thinking has not influenced most physicists. Indeed, I was thinking about laws evolving before I read Charles Sanders Peirce. But something he said encapsulates what I think is a very important conclusion that I came to through a painful route—and other people have more recently come to it—which is that the only way to explain how the laws of nature might have been selected is if there's a dynamical process by which laws can change and evolve in time. And so I've been searching to try to identify and make hypotheses about that process where the laws must have changed and evolved in time, because the situation we're in is: Either we become kind of mystics—"Well, those are just the laws," full stop—or we have to explain the laws. And if we want to explain the laws, there needs to be some history, some process of evolution, some dynamics by which laws change.

This is for some people a very surprising idea, and it still is a surprising idea in spite of the fact that I've been thinking about it since the late '80s, but if you look back, there are precedents: Dirac. You can find in his writings a place where Dirac says the laws must have been different earlier in the universe than now; they must have changed. Even Feynman has. . . . I found a video online where Feynman has a great way . . . and I wish I could do a Feynman Brooklyn accent. It sort of goes: "Here are the laws, we say; here are the laws, but how do they get to be that way in time? Maybe physics really has a historical component." Because, you see, he's saying physics is different from the other subjects. There's no historical component to physics, as there is to biology, genealogy, astrophysics, and so forth. But Feynman ends up saying, "Maybe there's a historical component." And then in the conversation his interviewer says, "But how do you do it?" And Feynman goes, "Oh, no, it's much too hard, I can't think about that."

So having said that, it's very audacious to say I've been trying to think about that since the late '80s.

Lee Smolin

It's worth mentioning what got me started thinking about evolving laws, and that was a comment that my friend Andy Strominger made about string theory. Andy is one of the important string theorists in the United States. Andy had just written a paper, I think in about '88, in which he had uncovered evidence for the existence of a vast number of string theories. So originally there were five, and maybe that was not so bad; they could be unified. And then there were hundreds of ways, and then there were hundreds of thousands of ways, to curl up the extra dimensions. And then Andy identified another way to make a string theory that would make vast numbers. And he said to me, "It's not even going to be worthwhile trying to connect this theory to experiment, because whatever comes out of an experiment, there is going to be a version that would match it."

And it took a lot of people a long time, until the early 2000s, to catch up to that. But I was really struck by that conversation and then went away and wondered about this: How could you have a theory that accounts for the selection of laws from a vast catalogue of possible laws? And not only that, there are some mysteries about why the laws are what they are—because they seem to be very special in certain ways. One way they're very special is that they seem to be chosen in such a way that it leads to a universe with an enormous amount of structure. With structure on every scale, from molecules and biological molecules, to biological systems themselves, to all the rich variety of structures on the Earth and the other planets, to the rich structures of galaxies, to clusters of galaxies on this vast array of scales.

The universe is not boring on any scale you look at it. It's very structured. Why? And there turned out to be two connected reasons. One of them is that the laws are very special. One way they're very special is that they have parameters in them, which take values that we don't know the reasons for. These are things like the masses of the different elementary particles—the electrons, the neutrinos, the quarks—and the strengths of the fundamental forces. I'm talking about thirty numbers that we just put into theory from experiment. And then we have a model—the standard model—which works very well. But we don't understand why those numbers are what they are. So I started to imagine a scenario where the numbers could change in some violent events. Maybe the Big Bang was not the first moment in time; maybe it was a violent event where our universe grew out of some previous universe, and maybe those numbers altered the way that when a new individual is born the genes are different from the parents'.

And I started to play with that idea and began to see how you could use the principles of natural selection to make predictions about our present universe. These predictions test the scenario that the laws have evolved in a particular way. A thing I understood from that: There was already speculation about multi-universes and our universe being one of a vast number of other universes, and there was already the use of the anthropic principle to pick out our world. But I realized you can't do science assuming that our universe is one of a vast array of other universes, because we can't observe any properties of them. And I've been making this argument forever, and it doesn't seem to penetrate to some people, that science is not a fantasy story. It's not a Harry Potter story about magical things that might be true. Science is about what you can verify—hypotheses that you can test and verify. If you're making hypotheses about many universes that exist simultaneously with us with no connection to our own, you can't verify those hypotheses. But if you're making hypotheses about how our universe evolved from past universes, you're making

Lee Smolin

hypotheses about things that happened in our past and there can be consequences that you can verify. So through this, I came to the idea that laws must have evolved in time. And that was the idea of cosmology and natural selection.

Now, meanwhile most of my work has been about making the quantum theory of gravity. And in quantum gravity, we apply quantum mechanics to the equations of Einstein's general theory of relativity and we come up with a theory that has no time, fundamentally. This is a point that Stephen Hawking made, that Julian Barber has made, in many different ways. The variable time—the dependence of processes on time—just disappears from the fundamental equations of quantum cosmology, of quantum gravity applied to cosmology. And time is set to emerge when the universe gets big, in the same way the temperature emerges as an approximate description of the energy contained in a lot of molecules moving around randomly—where pressure emerges as a summation of all the forces coming from all the collisions of an atom on a wall.

But time is nowhere in the fundamental equations of quantum cosmology. And I was working on the equation of quantum cosmology for many years, first with Ted Jacobson and then with Carlo Rovelli. We solved a form of those equations, and that was the main root of my work in quantum gravity. So for many years I had these two parallel things going on: one in which laws were evolving in time and the other in which time was emerging from laws, which therefore implied that the laws were timeless.

And because the first thing was a kind of side project—it was a kind of thing I had thought about occasionally, on the side of my main work—it took me a long time to realize that there was a contradiction between those two stories. I'm a little bit ashamed about that, but it's better to lay it on the table. And several things happened which made that contradiction very evident and made me deal with it. One of them was in the quantum theory of gravity itself. There turned out to be technical issues realizing the picture of time emerging from a timeless quantum cosmology. This isn't the place to talk about technical issues, but something I'm convinced about is when a technical issue hangs around for many years and many people work on it and nobody solves it, it may be that you should reexamine the ideas behind it. Maybe it's not a technical issue. Maybe it's a fundamental conceptual or philosophical issue.

And indeed this is something that Feynman said to me when I was a graduate student. He said there are things—again, I'm not sure why I'm invoking Feynman so much, but why not? He said, "There are things that everybody believes, that nobody can demonstrate. And you can make a useless career in science . . ." He probably put it in an even harder way, "You can waste your time and waste your career by trying to work on things that everybody believes but nobody can show, because you're probably not going to be able to show them either, if a lot of smart people couldn't show them. Or you can investigate the alternative hypothesis, which is that everybody's wrong." And this always comes back to me, this, with Feynman's voice. He, at the time, was thinking the confinement in QCD was wrong, and he was probably wrong about that, but nonetheless he made a brave effort to prove confinement.

So I began to wonder whether we might be wrong about time emerging from law in quantum cosmology. I began to think maybe we should make quantum cosmology in some way in which time is fundamental and space may emerge from something more fundamental. So that's one thing that happened.

The second thing that happened is that the picture of laws evolving, of a collection of universes evolving on a landscape of laws, went in about 2003 from being very much kind of a one-person little obsession that I engaged in on the side to a big deal when a bunch of people in string theory came to the same conclusion. This was a result of work at Stanford, and the impetus for this was making string theory accommodate itself to the positive dark energy, or positive cosmological constant or vacuum energy.

And this collaboration of people at Stanford discovered that they could make string theories that had positive vacuum energy but only at the cost of there being vast numbers of string theories. So, really, they got back to where Andy Strominger was in 1988. And all of a sudden—and Lenny Susskind here played a big role—all of a sudden there was a lot of talk about the "landscape of theories" and the "dynamics" and "change" on the landscape of theories, which were the words I had used. So I was sort of jolted into, "My god, if everybody is taking this seriously, I should think carefully about this."

The third thing that happened was that I started interacting with a philosopher, Roberto Mangabeira Unger, who had on his own been thinking about evolving laws for his own reasons. And he basically took me to task—this is about six or seven years ago—and said, "Look, you've been writing and thinking about laws evolving in time, but you haven't thought deeply about what that means for our understanding of time. If laws can evolve in time, then time must be fundamental." And I said "Yes," but he said, "You haven't thought deeply, you haven't thought seriously about that." And we began talking and working together as a result of that conversation.

And so those three things together, about five or six years ago, made me go back and put together the idea that laws have to change in time if they're to be explained—with my thinking about the nature of space and time, quantum mechanically, and so I started playing with the idea that maybe time has to be really fundamental in the context of quantum gravity.

That thinking changed my work, and much of my work for the last year has been devoted to thinking about various ways, various hypotheses, about how laws can change in time. Thinking about the consequences for understanding the nature of time, and thinking about how to make theories and hypotheses that can be checked. The reason is that this stuff can get pretty speculative. I'm sure it does sound speculative—so to tie it down, I focus on hypotheses that are testable.

Feynman once told me, "You're going to have to do crazy things to think about quantum gravity. But whatever you do, think about nature. If you think about the properties of a mathematical equation, you're doing mathematics and you're not going to get back to nature. Whatever you do, have a question that an experiment could resolve at the front of your thinking." So I always try to do that.

Let me first mention that cosmological natural selection did make some predictions, and those predictions have so far stood up. Let me talk about some newer ideas.

Let me give some more examples, because cosmological natural selection was a long time ago. Here's one that I call the principle of precedence. And I think it's kind of cute. And let me phrase it in the language of quantum mechanics, which is where it comes from. It comes from actually thinking about the foundations of quantum mechanics, which is another thing I try to think about occasionally. We take a quantum system, and quantum systems are always thought about, from my point of view, as small bits of the universe that we manipulate and prepare in states

Lee Smolin

and experiment with and measure. We're always *doing* something to a quantum system. I don't believe anymore that there's anything that goes under the name of "quantum cosmology."

Let's say we have a quantum system—let's say some ions in an ion trap, and we want to measure their quantum mechanical properties. So we prepare them in some initial state. We evolve them by transforming them, by interacting with them from the outside, for example, by applying magnetic fields or electric fields or probing them with various probes. And then we apply a measurement. And because it's quantum mechanics, there's no prediction for the definite outcome of the experiment; there are probabilities for different possible outcomes.

Let's consider a system that's been studied many times. We have measured the statistical distribution of outcomes through some collection of past instances, where we've measured the system before. And if we do it now and measure the system again, we're going to get one of those past outcomes that we saw before. If we do it many times, we're going to get a statistical distribution, which is going to be the same distribution that we saw before. We're confident that if we do it next year, or in a million years, or in a billion years, we're going to get the same distribution as we got before. Why are we confident of that? We're confident of that because we have a kind of metaphysical belief that there are laws of nature that are outside time, and those laws of nature are causing the outcome of the experiment to be what it is, and laws of nature don't change in time, they're outside of time. They act on the system now, they acted on the system in the same way in the past, they will act the same way in a year, or a million or a billion years, and so they'll give the same outcome. So nature will repeat itself, and experiments will be repeatable because there are timeless laws of nature.

But that's a really weird idea, if you think about it, because it involves the kind of mystical and metaphysical notion of something that is not physical, something that is not part of the state of the world—something that is not changeable acting from outside the system to cause things to happen. And when I think about it, that is kind of a remnant of religion. It's a remnant of the idea that God is outside the system, acting on it.

So let's try a different kind of hypothesis. What if, when you prepare the system, transform it, and then measure it, nature has a way of looking back and asking, "Have similar things been done in the past? If they have, let's take one of those instances randomly and just repeat it." That is, nature forms habits. Nature looks to see whether a similar thing happened in the past. And if there was, what if it takes that? If there are many, it picks randomly among them and presents you with that outcome. OK, well, that will give the same statistical distribution you saw in the past, by definition, because you're sampling from the past. So there doesn't have to be a law outside of time. The only law needs to be what I call the principle of precedence—that when you do an experiment, nature looks back and gives you what it did before.

Now, you can say that that involves some weird metaphysical idea that nature has access to its past and is able to identify when a similar thing is being done, a similar measure being made. And that's true, but it's a different metaphysical idea from the idea of the law acting from the outside, and it has different consequences. Let's play with this. This reproduces the predictions of standard quantum mechanics, so it reproduces the success of standard quantum mechanics without having to believe in the timeless law.

Can you test it? As I said, I'm only interested in ideas that can be tested. Yes, you can test it, because people are working a lot

Lee Smolin

with quantum technologies and they're making systems that have no precedents. For example, in Waterloo, there's the Institute for Quantum Computing, and there Ray Laflamme and David Cory and their colleagues are making systems that have never been made before. And so I talk to them and I say, "Maybe if you make a really novel system, then we'll have no precedents. It won't behave as you expect it to, because it won't know what to do, so it will just give you some totally random outcome." And they laugh, and I say, "Why is that funny?" And they say, "Well, the first time we do an experiment, of course we get a totally random outcome, because there are experimental-design issues and experimental-error issues. We never get what we think we're going to get, the first time we set up an experiment in the laboratory and we run it." So I say to them, "Great! That's fine. But eventually the thing settles down and starts giving repeatable results?" and they say, "Sure." And I say, "Well, could you separate that process"—of settling down to definite results—"could you separate out the effect of having to make your experiment work from the effect of my hypothesis that nature is developing habits as it goes along?" And they say, "Maybe." So we have a discussion going on, about whether this can really be tested.

Now, it's like any idea—it's probably wrong but it's testable, and that, to me, proves that it's science.

The current scene is very confusing. Very smart people have tried to advance theoretical physics in the last decades, and we're in an embarrassing situation. The embarrassing situation is that theories that were already around in the middle '70s for particle physics and the early '80s for cosmology are being confirmed over and over again, and to greater and greater precision, by the current experiments. And this goes both for particle physics and cosmology.

In particle physics, the LHC, the Large Hadron Collider, identified a new particle that is probably the Higgs—it looks like a standard-model Higgs—and there's *nothing else*. There's no evidence for supersymmetry, for extra dimensions, for new generations of quarks, for substructure—for a whole variety of ideas, some of which have been very popular and some of which have not been very popular but nonetheless have been on the table. None of these ideas that go beyond the standard model have been confirmed. In cosmology, the results of the *Planck* seem to be right in line with the simplest version of inflation. And this is a triumph for the standard model and a triumph for inflation.

Paul Steinhardt has a very interesting argument that the results of *Planck* really shouldn't be taken as confirming inflation. I have enormous respect, I have deep respect, for Paul, but the case is not closed. The case is not closed. And certainly, at a naïve level, it looks like a universe that's exactly the one that the inflationary theorists told us about, and they should be proud of that. But the situation leaves a conundrum, because we have nothing that confirms anything that goes beyond those models.

And I should say that there has been, in my field of quantum gravity, a lot of interest in the idea that certain astrophysical experiments would be able to see the breakdown of the structure of space and time that we have from general relativity, and give us evidence of quantum space and time in the propagation of light coming from faraway gamma-ray bursts, in the propagation of cosmic rays. We've been expecting for about a decade to see signals of quantum spacetime. And those are not there either, so far. So we're also very frustrated.

So my impression is that when— Let me just come back to how I quoted Feynman: When very smart people have been working under certain assumptions for a long time—and these ideas have been around for a lot longer than the ideas that Feynman was concerned with were—and we're not succeeding in uncovering new phenomena and new explanations, new understanding for phenomena, it's time to reassess the foundations of our thinking. It doesn't mean that everybody should do that, but some people should do that. And I find myself doing that because of my own intellectual history, partly because of my work in quantum gravity, partly because of cosmological natural selection, partly because I have an inclination—because part of my education was in philosophy. Although my PhD is in physics, part of my undergraduate was in philosophy, and I've always had an interest in philosophy. And I've always had even more than that—an appreciation, a deep appreciation, for the history of thought about these fundamental questions. I find myself doing some of this reassessment. And we'll see where it goes.

The conclusions that I come to, I think they're not subtle, they're easy to list, are, first, that—and I was opening with them before—the method of physics with fixed laws which are given for all time, acting on fixed spaces of states which are given for all time, is self-limiting. The picture of atoms with timeless properties moving around in a void according to timeless laws, this is self-limiting. It's the right thing to do when we're discussing small parts of the universe, but it breaks down when you apply it to the whole universe or when your chain of explanation gets too deep.

Let me give one reason it breaks down. We can use the language of reductionism. It's very good advice, it has worked for hundreds of years, that if we want to understand the properties of some composite system, some material, we explain it in terms of the properties of its parts or the things it's made from. That's good common sense, and a lot of the success of science is due to applying that good commonsense advice.

But what happens when you get to the things you think of as the elementary particles? They have properties, too. They have masses and charges, with various forces that they move about with. But they have no parts, we believe. Or if they do have parts, you're just continuing to do this, and then you should be looking for the breakdown into the parts that experiment has not seen so far.

Is there any other way to explain the properties of fundamental particles? Well, not by further reductionism. There has to be a new methodology. So that's the first conclusion: that the methodology that works for physics and has worked for hundreds of years, there's nothing wrong with it in the context in which it's been applied successfully, but it breaks down. When you push to the limits of explanation, reductionism breaks down. It also breaks down when you push on the other end, to larger and larger systems, to the universe as a whole. I mentioned several reasons why it breaks down but there are others. Let me mention one: When we experiment with small parts of the universe, we do experiments over and over again. That's part of the scientific method; you have to reproduce the results of an experiment, so you have to do it over and over again. And by doing that, you separate the effect of general laws from the effect of changing the initial conditions. You can start the experiment off in different ways and look for phenomena which are still general. These have to do with general laws. And so you can cleanly separate the role of initial conditions from the role of the general laws.

When it comes to the universe as a whole, we can't do that. There's one universe, and it runs one time. We can't set it up. We didn't start it. And indeed, in working cosmology, in inflationary theory, there's a big issue, because you can't separate testing hypotheses about the laws from testing hypotheses about the ini-

Lee Smolin

tial conditions—because there was just one initial condition and we're living in its wake. This is another way in which this general method breaks down. So we need a new methodology.

A good place to look for that methodology is in the relational tradition, the tradition of Leibniz and Mach and Einstein, that space and time and properties of elementary particles are not intrinsic but have to do with relationships that develop dynamically in time. This is the second point.

The third conclusion is that time therefore must be fundamental. Time must go all the way down. It must not be emergent, it must not be an approximate phenomenon, it must not be an illusion. These are the conclusions that I come to and that my work these days is based on.

So how do I situate myself? There are two areas that my work impinges on most directly. One of them is quantum gravity and the other is cosmology. Let me discuss each of those in turn. In quantum gravity, there are several programs of research. The one I'm most identified with is loop quantum gravity. Loop quantum gravity is doing very well, and let me take a minute for that, because we haven't talked about that.

Loop quantum gravity is a conservative research program. It comes from applying quantum mechanics directly to a form of general relativity, with no additional hypotheses about extra dimensions or extra particles or extra degrees of freedom. The particular form of general relativity we use is very close to gauge theory. It's very close to Yang-Mills theory. This was a form developed by Abhay Ashtekar, and before him, although we didn't know about this at the time, by Plebanski. This is now a big research program.

We have every two years an international conference. This year I'm among the organizers. We're doing it at Perimeter, and we already have—it's very early, the conference is not until July—and we already have more than two hundred people registered to come. So this is not Carlo Rovelli and Abhay Ashtekar and me sitting around in Verona writing in our notebooks, the way it was in the late '80s—which, by the way, was a great experience. It's great to be an inventor of something, and it's great to have a period like that.

Loop quantum gravity gives us a microscopic picture of the structure of quantum geometry with the Planck scale, which is 20 orders of magnitude smaller than an atom. The key problem that loop quantum gravity has had to face is, How does the space-time we see around us emerge from that quantum picture? How do the equations of general relativity emerge to describe the dynamics of that spacetime on a big, macroscopic scale? And there's been a lot of progress to answering those questions in the last five or ten years. So it's very healthy as a theoretical research program.

However, there are two big frustrations with it. One of them is that it still doesn't connect to experiment. I and others have been hoping we would be able to make measurements that would detect the quantum structure of the geometry of spacetime. Those are astrophysical experiments, and those experiments are not showing any sign of that quantum structure. And the other thing is that from my present point of view, loop quantum gravity is successful also when applied to small parts of the universe, but I no longer believe in taking equations of quantum gravity and applying them to the universe as a whole, because time disappears when you do that and I think that time is fundamental. But loop quantum gravity is healthy and is making the kind of incremental progress that healthy research programs make—which doesn't mean it's right, but it means that it's solving the questions it has to solve to be real science.

Lee Smolin

There are amazing young people working in the subject, people who are technically brilliant, who are able to do things that just amaze me, and that's a great pleasure. I'm partly in and partly not in that research community, because my interests in cosmology and my interest in the nature of time and other interests take me outside of it. But I still have many good friends there; I still go to the conferences. Some of what I do is in that context of loop quantum gravity, and I'm very happy to be a part of that community. But I'm not sitting at the center of it anymore, which is fine, because the people who are sitting at the center of it are better able to hold that position than I am.

String theory, which I've also worked on, is in part healthy as a research program and in part stuck. We no longer hear much from string theorists about what is the fundamental formulation of string theory, or M-theory, as we used to call it, which is the part I was most interested in and tried to work on. And we no longer hear—although I think many people still believe them—we no longer hear aggressive claims of string theory being a theory of everything.

There are two areas in which string theory is doing very well. One of them is mathematically. It's beautiful mathematics and mathematical physics. And it also provides applications—through what's called the Maldacena conjecture, or the AdS/CFT conjecture, to give the jargon—to ordinary systems: liquids, fluids, certain solid-state systems. The same methods can be used to in a new way illuminate some experimental phenomena. That has nothing to do with string theory as a unified theory, but it's developing very nicely. Then there are other programs. There is causal dynamical triangulations, quantum gravity, causal sets— these are things worked on by handfuls of people, and they are part of the landscape of ideas.

In cosmology, inflation and the standard model of cosmology are doing very well observationally. But I think that Neil Turok and Paul Steinhardt have a very important point, which I agree with, which is that if you don't address the singularity, the time before inflation—if inflation is true—when the universe becomes infinitely dense at a finite time in the past and general relativity stops working, you can't address, really, the question of what chose the initial conditions. And also, to me, what chose the laws.

It seems to me a necessary hypothesis that the Big Bang was not the first moment of time but was an event—a transition, something like a phase transition, before which there was a universe that had possibly different properties and different laws. So the Big Bang becomes a phase transition, something like a black hole that formed in a previous universe. There would have been a singularity to the future of that formation of that black hole and instead that singularity is wiped out by quantum effects and, as we say, bounces. Whereas the star was collapsing and was going to just collapse to infinite density, quantum effects make it bounce back and start expanding again. And that makes a new region of space and time, which can be a new universe.

That's one hypothesis about what the Big Bang was as a transition. Paul and Neil have a different hypothesis, which has to do with the whole universe, as a whole, going through a phase transition. The quantum-gravity people have a different hypothesis. According to this hypothesis, it's as if the properties of space as we know it are like a frozen piece of ice, and when the universe goes through a Big Bang it's like space melts and becomes liquid and rearranges its properties and then freezes again. The Big Bang was something like a big freeze following a temporary melting. It seems to me that it's a necessary hypothesis to explain the initial conditions, because the one thing that inflation doesn't do, which

Lee Smolin

it claimed to do, was make the initial conditions of the universe probable or explain why the universe is so unusual in its early stages. And whatever the fate of Paul and Neil's cyclic cosmology, their particular hypothesis, I think they're right in their critique of inflation. Whether inflation is right or not, I think they're right that there had to have been a phase transition replacing the Big Bang and therefore the explanations for things in the early universe will be pushed back to before the Big Bang.

And that, of course, intersects with my interest in quantum gravity, because quantum mechanics has to become important at those scales where the phase transition happened. And, indeed, over in the quantum gravity world we have models of quantum cosmology, so-called loop-quantum-cosmology models, developed by Martin Bojowald, Abhay Ashtekar, and many other people by now, which show this bounce happening—show that the singularities are always removed and replaced by bounces.

Cosmology has been very healthy because of the success of the standard model of cosmology. But we're left with a question similar to the question the particle physicists are left with: the question of Why this peculiar universe? We've measured the properties of the universe very well. Whether inflation is true or not, it's an improbable universe. Why this universe? Why not other universes that would be more typical, given what we understand about the laws?

This is the initial-conditions problem. One of my major points is that we can't address those kinds of questions on the basis of the same methodology that has worked so far. We need a new methodology for physics—one in which laws evolve.

Is there a community of people thinking the same way? Yes and no. There aren't very many people within either the cosmology community or the quantum gravity community. So, for example, Carlo Rovelli is my dear friend, and in some areas, like loop quantum gravity, we are very much in sync, but Carlo is still a believer in the fundamental timelessness of quantum gravity and quantum cosmology and I am not. Although we talk about it.

In the world of philosophy, what I'm doing is not new and not a surprise. I mentioned Roberto Unger. Our collaboration has been at times like what Picasso once said about his collaboration with Braque: It's like being at times roped together on a mountain. And with Roberto, it's been a wonderful adventure, to develop these ideas and to provoke each other. There's also a philosophical context going back to the American pragmatist tradition, going back to Charles Sanders Peirce. In that context, none of the ideas I'm talking about are new or particularly surprising. So, how philosophers will react is unclear, but I'm in a context—the context of ideas I'm talking about, in which time is real and laws can change—these are issues that they've been talking about and discussing and debating and have positions on already for a century.

I hope to convince people, because the chain of thought that I've been through is not serendipity, it's not where I planned to end up, and it's not where I hope to end up. I don't actually like being out on my own. I don't actually enjoy controversy and conflict, unlike some other people we can mention. I feel like my job is to develop these ideas, to put them out there, and especially to develop them in a form in which they're science and not philosophy. The philosophers can develop the philosophy.

And let me mention, if I may, another ramification of the idea that time is real as opposed to emergent or an illusion. The second law of thermodynamics is very well established and is, on a microscopic scale, clearly true. Disorder increases, entropy increases, most things we deal with in our everyday lives are irreversible. There are strong arrows of time. There's a directionality of time: We can't go backwards. We are born, we grow up, we get older, and we die. If I spill this Coke on the carpet, nothing we can do can make it go back into the cup. The birth of a child is irreversible. An unkind word said accidently to a friend is irreversible. Many things, most things, in life are irreversible. This is mostly codified by the second law of thermodynamics.

In the late 19th century, Boltzmann proposed successfully that thermodynamics was not fundamental, because matter was made of atoms. Instead, he proposed that the laws of thermodynamics could be explained as being emergent from the behavior of atoms, so that they are a consequence of the fundamental laws the atoms obey. So temperature is not a fundamental quantity, it's the average energy caught up in the random motion of atoms, and so forth. And entropy is not a fundamental quantity, it's a measure of the disorder or the improbability or the probability of a configuration of atoms.

Boltzmann was right, but there was a paradox inherent in his reasoning which people at the time identified. It's shocking to think that, but at the time, in the late 19th century, the atomic hypothesis was not wildly popular, and there was no consensus among physicists. So he had intellectual opponents. And they said to him, "You claim to have derived as emergent a theory that has a strong directionality of time, from the fundamental laws of motion of Newton. But the fundamental laws of motion of Newton are reversible in time." If you take a picture, take a film of atoms moving about in a void, interacting according to Newton's laws, and you run that film backwards, that's something that also can happen, according to the laws. So there's a kind of paradox, because Boltzmann could just as easily have used Newton's laws to prove the anti-second law, to prove that entropy is always higher in the past and lower in the future.

And, indeed, the critics were right. The right way to resolve this puzzle was worked out by the Ehrenfests—Paul and Tatyana Ehrenfest, who were dear friends of Einstein around 1905, 1908, I think, or somewhere around then. And they understood that actually what Boltzmann had proved was symmetric in time. What he proved is that if you find a system with the entropy low at one time, it's most likely to increase in the future, because disorder most likely increases when things move about randomly. But it's *also* most likely that the entropy was higher in the past and that what you're seeing is an accident—what he called a fluctuation. And so the question is not "What explains the second law?" but "What explains the conditions, the initial conditions?"

To explain the second law, you have to assume that the initial conditions are improvable, so the system starts out more ordered than it might be. And this was a great mystery to Boltzmann. He didn't have the benefit of living long into the 20th century, so he imagined that the universe was governed by Newton's law and was eternal. And he could only assume that we lived in the wake of a huge fluctuation where the universe was mostly in equilibrium—which is the state when entropy is maximal—and spent most of its time in equilibrium and just occasionally, due to a random fluctuation, got way out of equilibrium. And that formed the sun, and that was the cause of the world we were living in now. Now, that's wrong. There's no evidence for that.

So why is there such a strong arrow of time, if the laws of physics are fundamentally reversible in time? Well, Roger Penrose had an idea I think is very worth investigation, which is in two parts. One of them—this was in an essay in 1979—he argued, and I think correctly, that the only way to explain the arrow of time as we observe it in the universe is if the initial conditions of the Big Bang were very, very special and very, very improbable. And that's a theme of my discussion, and that reoccurs here.

So, yet another sense in which to explain, within the present paradigm of reversible laws the tremendous irreversibility of phenomenon that we observe, you have to put all the weight on the cosmologists to explain why the initial conditions were so improbable. And, as a cosmologist I know remarked, that's not a job the cosmologists signed up for. They have enough to do. They have enough problems of their own, let alone having to explain the whole irreversibility of nature and the second law. But that's where the burden of proof is.

Now, Roger Penrose's proposal was that maybe the fundamental laws are actually time-asymmetric and the time-symmetric laws are emergent and approximate. And so maybe all those histories where we take a movie of part of the universe and run it backwards couldn't really be part of history of the real universe going all the way back to the Big Bang. Let me give an example of that.

When we look around, we see light coming from the past. I mean it's more evident when we look out in telescopes; we see stars as they were in the past. We never see light coming to us from the future. We never see starlight coming to us from stars in the future. We never see supernova explosions in the future sending radiation back in time to us. But the laws that govern the propagation of light, Maxwell's equations, are reversible in time, so it has solutions that involve light propagating from events in the future and propagating information and energy into the past for us to observe from the past. It has just as many solutions like that as it has solutions of the kind we use. So the law is symmetric in time, but to apply it to nature we throw away most of it, because we throw away any solutions where there's any hint of anything propagating from the future to the past.

Roger would say, "Maybe the real theory that underlies Maxwell's equations"—which for him, and I agree, would be the real quantum theory of gravity—"just propagates energy and information from the past into the future and doesn't have this problem and this paradox." So this then becomes a challenge. Can we make hypotheses about how the fundamental laws could really be asymmetric in time and irreversible in time and understand how the present laws become reversible? And with a colleague, Marina Cortes, I've been working on that.

So that's another way in which these philosophical critiques that I think are necessary to understand why we're stuck in fundamental physics and cosmology serve to motivate my work as a scientist. That work is then to be appreciated or not and evaluated on the basis of the usual criteria of science. That is, does it lead to new hypotheses that lead to new experiments to check them?

10
The Landscape

Leonard Susskind

Felix Bloch Professor of theoretical physics, Stanford University; coauthor (with Art Friedman), *Quantum Mechanics: The Theoretical Minimum*

INTRODUCTION by John Brockman

For some people, the universe is eternal. For me, it's breaking news.

Back in 2003 I sat down to talk with Lenny Susskind, the discoverer of string theory. After he left, I realized I had become so caught up in his storytelling that I forgot to ask him, "What's new in the universe?" So I sent him an email. Here's his response:

The beginning of the 21st century is a watershed in modern science, a time that will forever change our understanding of the universe. Something is happening which is far more than the discovery of new facts or new equations. This is one of those rare moments when our entire outlook, our framework for thinking, and the whole epistemology of physics and cosmology are suddenly undergoing real upheaval. The narrow 20th-century view of a unique universe, about 10 billion years old and 10 billion light years across with a unique set of physical laws, is giving way to something far bigger and pregnant with new possibilities.

Gradually physicists and cosmologists are coming to see our ten billion light years as an infinitesimal pocket of a stupendous megaverse. At the same time theoretical physicists are proposing theories which demote our ordinary laws of nature to a tiny corner of a gigantic landscape of mathematical possibilities.

This landscape of possibilities is a mathematical space representing all of the possible environments that theory allows. Each possible environment has its own laws of physics, elementary particles and constants of nature. Some environments are similar to our own corner of the landscape but slightly different. They may have electrons, quarks and all the usual particles, but gravity might be a billion times stronger. Others have gravity like ours but electrons that are heavier than atomic nuclei. Others may resemble our world except for a violent repulsive force (called the cosmological constant) that tears apart atoms, molecules and even galaxies. Not even the dimensionality of space is sacred. Regions of the landscape describe worlds of 5,6 . . . 11 dimensions. The old 20th-century question, "What can you find in the universe?" is giving way to "What can you not find?"

The diversity of the landscape is paralleled by a corresponding diversity in ordinary space. Our best theory of cosmology, called inflationary cosmology, is leading us, sometimes unwillingly, to a concept of a megaverse, filled with what Alan Guth, the father of inflation, calls "pocket universes." Some pockets are small and never get big. Others are big like ours but totally empty. And each lies in its own little valley of the landscape.

Man's place in the universe is also being reexamined and challenged. A megaverse that diverse is unlikely to be able to support intelligent life in any but a tiny fraction of its expanse. Many of the questions that we are used to asking, such as "Why is a certain constant of nature one number instead of another?" will have very different answers than what physicists had hoped for. No unique value will be picked out by mathematical consistency, because the landscape permits an enormous variety of possible values. Instead the answer will be "Somewhere in the megaverse the constant is this number, and somewhere else it is that. And we live in one tiny pocket where the value of the constant is consistent with our kind of life. That's it! There is no other answer to that question."

The kind of answer that this or that is true because if it were not true there would be nobody to ask the question is called the anthropic prin-

ciple. Most physicists hate the anthropic principle. It is said to represent surrender, a giving up of the noble quest for answers. But because of unprecedented new developments in physics, astronomy, and cosmology these same physicists are being forced to reevaluate their prejudices about anthropic reasoning. There are four principal developments driving this sea change. Two come from theoretical physics, and two are experimental or observational.

On the theoretical side, an outgrowth of inflationary theory called eternal inflation is demanding that the world be a megaverse full of pocket universes that have bubbled up out of inflating space like bubbles in an uncorked bottle of Champagne. At the same time, string theory, our best hope for a unified theory, is producing a landscape of enormous proportions. The best estimates of theorists are that 10^{500} distinct kinds of environments are possible.

Very recent astronomical discoveries exactly parallel the theoretical advances. The newest astronomical data about the size and shape of the universe convincingly confirm that inflation is the right theory of the early universe. There is very little doubt that our universe is embedded in a vastly bigger megaverse.

But the biggest news is that in our pocket the notorious cosmological constant is not quite zero, as it was thought to be. This is a cataclysm and the only way that we know how to make any sense of it is through the reviled and despised anthropic principle.

I don't know what strange and unimaginable twists our view of the universe will undergo while exploring the vastness of the landscape. But I would bet that at the turn of the 22nd century, philosophers and physicists will look back nostalgically at the present and recall a golden age in which the narrow provincial 20th century concept of the universe gave way to a bigger better megaverse, populating a landscape of mind-boggling proportions.

Below is a wide-ranging Edge conversation with Lenny in December 2003 on the anthropic principle and on the early history of string theory:

The Landscape

What I mostly think about is how the world got to be the way it is. There are a lot of puzzles in physics. Some of them are very, very deep, some of them are very, very strange, and I want to understand them. I want to understand what makes the world tick. Einstein said he wanted to know what was on God's mind when he made the world. I don't think he was a religious man, but I know what he means.

The thing right now that I want to understand is why the universe was made in such a way as to be just right for people to live in it. This is a very strange story. The question is why certain quantities that go into our physical laws of nature are exactly what they are, and if this is just an accident. Is it an accident that they are finely tuned, precisely, sometimes on a knife's edge, just so that the world could accommodate us?

For example, there's a constant in nature called the cosmological constant, and it's a certain number. If that number differed by the tiniest amount from what it really is, the universe could not have been born with galaxies, stars, planets, and so forth. Is it an accident that the number was exactly right to be able to form the universe as we see it? Or is it some feature of the way the universe works that makes it necessarily create life? It sounds crazy and most physicists think such thoughts are hogwash, but I'll give you an example.

Suppose we lived on a planet and we couldn't see out because there was too much fog and too many clouds. Suppose we wanted to know why the temperature on this planet is precisely right for us to be able to live without getting cooked and without getting frozen. Is it an accident, or is there a design involved? Most people, knowing the answer, would say that if you look out far away into the cosmos, you see all kinds of planets, stars, empty regions and so forth. Some of them are much too hot to live on, some of them are much too cold to live on, and some of them are in between but don't have water. There are all kinds of planets are out there.

The answer is that we simply live on the planet that we can live on because the conditions are exactly right. It's an environmental fact that conditions are exactly right, so it's no accident that we happen to find ourselves in an environment which is finely tuned, and which is precisely made so that we can live in it. It's not that there's any law of nature that says that every planet has to be livable, it's just that there are so many different things out there—roughly 10^{22} planets in the known universe, which is a huge number—and surely among them there will be a small number which will be at the right temperature, the right pressure, and will have enough water, and so forth. And that's where we live. We can't live anywhere else.

The question is whether our environment in a bigger sense—in terms of the laws of nature that we have, the elementary particles, the forces between them, and all those kinds of things—are environmental things which are contingent in our particular region of the universe or are exactly the same throughout the whole universe. If they're contingent, that means they may vary from place to place, or they may vary from one thing to another thing to another thing. If that were the case, then we would answer some subset of the questions we're interested in by saying things are the way they are because if they were any other way we couldn't live here. The environment has to be right for us to exist.

On the other hand, if everything is the same, all across the universe from beginning to end, then we *don't* understand why things are tuned in the way that allows us, with knife-edge precision, to be in an environment that supports life. This is a big controversy that's beginning to brew in physics: whether the laws of nature as we know them are simply derivable from some mathematical theory and could not be any other way or if they might vary from place to place. This is the question I would like to know the answer to.

In the United States, the cosmologists don't like the idea of the anthropic principle at all. In England they love it. I was very surprised to find out, when I started talking about this, that the physicists like myself, people who are interested in theoretical, mathematical questions in physics, are rather open to it in the United States, but the cosmologists are not. This [anthropic] idea originated to a large extent among British cosmologists—Martin Rees being one of them, John Barrow being another one. There's also Andrei Linde, who is a Russian but of course lives in the United States, who was one of them, as was Alexander Vilenkin. But that's not the crowd I'm addressing my remarks to.

The crowd I'm addressing are the high-energy physicists, the string theorists, and includes the Brian Greenes, the Ed Wittens, the David Grosses, and so forth. The reason is because over the last couple of years we've begun to find that string theory permits this incredible diversity of environments. It's a theory that simply has solutions which are so diverse that it's hard to imagine what picked one of them in the universe. More likely, the string theory universe is one with many different little patches of space that Alan Guth has called pocket universes. Of course they're big, but there are little patches of space with one environment, little patches of space with another environment, etc.

· Leonard Susskind

Mostly physicists have hated the idea of the anthropic principle; they all hoped that the constants of nature could be derived from the beautiful symmetry of some mathematical theory. And now what people like Joe Polchinski and me are telling them is that it's contingent on the environment. It's different over there, it's different over there, and you will never derive the fact that there's an electron, a proton, a neutron, whatever, with exactly the right properties. You will never derive it, because it's not true in other parts of the universe.

Physicists always wanted to believe that the answer was unique. Somehow there was something very special about the answer, but the myth of uniqueness is one that I think is a fool's errand. That is, some believe there is some very fundamental, powerful, simple theory which, when you understand it and solve its equations, will uniquely determine what the electron mass is, what the proton mass is, and what all the constants of nature are. If that were to be true, then every place would have to have exactly the same constants of nature. If there were some fundamental equation which, when you solved it, said that the world is exactly the way we see it, then it would be the same everywhere.

On the other hand, you could have a theory which permitted many different environments, and a theory which permitted many different environments would be one in which you would expect that it would vary from place to place. What we've discovered in the last several years is that string theory has an incredible diversity—a tremendous number of solutions—and allows different kinds of environments. A lot of the practitioners of this kind of mathematical theory have been in a state of denial about it. They didn't want to recognize it. They want to believe the universe is an elegant universe—and it's not so elegant. It's different over here. It's that over here. It's a Rube Goldberg machine over here. And this has created a sort of sense of denial about the facts about the theory. The theory is going to win, and physicists who are trying to deny what's going on are going to lose.

These people are all very serious people. David Gross, for example, is very harshly against this kind of view of diversity. He wants the world to be unique, and he wants string theorists to calculate everything and find out that the world is very special with unique properties that are all derivable from equations. David considers this anthropic idea to be giving up the hope for uniqueness, and he quotes Winston Churchill when he's with young people, and he says, "Nevah, nevah, nevah, nevah give up."

Ed Witten dislikes this idea intensely, but I'm told he's very nervous that it might be right. He's not happy about it, but I think he knows that things are going in that direction. Joe Polchinski, who is one of the really great physicists in the world, was one of the people who started this idea. In the context of string theory, he was one of the first to realize that all this diversity was there, and he's fully on board. Everybody at Stanford is going in this direction. I think Brian Greene is thinking about it. Brian moved to some extent from hardcore string theory into thinking about cosmology. He's a very good physicist. There were some ideas out there that Brian investigated and found that they didn't work. They were other kinds of ideas, not this diversity idea, and they didn't work. I don't know what he's up to now. I haven't spoken to him for all of a month. Paul Steinhardt hates the idea. Alan Guth is certainly very susceptible; he's the one who coined the term "pocket universes."

The reason there is so much diversity in string theory is because the theory has an enormous number of what I call moving parts, things you can tinker with. When you build yourself an example of string theory, as in Brian's book, it involves the geometry of these internal compact spaces that Brian became famous for studying. There are a lot of variables in fixing one of them, and a lot of variables to tinker around with. There are so many variables that this creates an enormous amount of diversity.

String theory started out, a long time ago, not as the theory of everything, the theory of quantum gravity, or the theory of gravitation. It started out as an attempt to understand hadrons. Hadrons are protons, neutrons, and mesons—mesons are the particles that fly back and forth between protons to make forces between them—just rather ordinary particles that are found in the laboratory that were being experimented on at that time.

There was a group of mathematically minded physicists who constructed a formula. It's a formula for something known as a scattering amplitude, which governs the probability for various things to happen when two particles collide. Physicists study particles in a rather stupid way; somebody described it as saying that if you want to find out what's inside a watch, you hit it as hard as you can with a hammer and see what comes flying out. That's what physicists do to see what's inside elementary particles. But you have to have some idea of how a certain structure of particles might manifest itself in the things that come flying out. And so in 1968 Gabrielli Veneziano, who was a very young physicist, concocted this mathematical formula that describes the likelihood for different things to come out in different directions when two particles collide. It was a mathematical formula that was just based on mathematical properties, with no physical picture, no idea of what this thing might be describing. It was just pure mathematical formula.

At that time, I was a very young professor in New York, and I was not an elementary-particle physicist. I tended to work on things like quantum optics and other things, just whatever I happened to be interested in. A fellow by the name of Hector Rubinstein came to visit me and my friend, Yakir Aharonov, and he was wildly excited. He said, "The whole thing is done! We've figured out everything!"

I said, "What are you talking about, Hector?"

And jumping up and down like a maniac, he finally wrote this formula on the blackboard. I looked at the formula and I said, "Gee, this thing is not so complicated. If that's all there is to it, I can figure out what this is. I don't have to worry about all the particle physics that everybody had ever done in the past. I can just say what this formula is in nice, little, simple mathematics."

I worked on it for a long time, fiddled around with it, and began to realize that it was describing what happens when two little loops of string come together, join, oscillate a little bit, and then go flying off. That's a physics problem that you can solve. You can solve exactly for the probabilities for different things to happen, and they exactly match what Veneziano had written down. This was incredibly exciting.

I felt, here I was, unique in the world, the only person to know this in the whole wide world! Of course, that lasted for two days. I then found that Yoichiro Nambu, a physicist at Chicago, had exactly the same idea, and that we had more or less by accident come on exactly the same idea on practically the same day. There was no string theory at that time. In fact, I didn't call them strings—I called them rubber bands.

I was just incredibly excited. I figured, "OK, here I am. I'm going to be a famous physicist. I'm going to be Einstein, I'm going to be Bohr, and everybody's going to pay great attention to me," so I wrote up the manuscript.

In those days we didn't have computers and we didn't have

email, so you hand-wrote your manuscript and gave it to a secretary. A secretary typed it, and then you went through the equations that the secretary had mauled and corrected them, and this would take two weeks to get a paper ready, even after all the research had been done and all you had to do was write it up. Then you put it in an envelope and you mailed it by snail mail to the editor of the *Physical Review Letters*. Now, the *Physical Review Letters* was a very pompous journal. They said they would only publish the very, very best. What usually happens when people start getting that kind of way is they wind up publishing the very worst, because when standards get very, very high like that, nobody wants to bother with them, so they just send it to someplace where it's easy to publish.

I sent it to the *Physical Review Letters*, and you understand, weeks had gone by in which I was preparing it and having it typed, and I was getting more and more nervous, thinking somebody was going to find out about it. I was telling my friends about it, and finally I sent the manuscript off. In those days it went to the journal, and the journal would have to mail it, again by snail mail, to referees. The referees might sit on it for a period and then send it back. All of this could take months—and it did take months.

And how did it came back? Well, they said, "This paper is not terribly important, and it doesn't predict any new experimental results, and I don't think it's publishable in the *Physical Review*."

Boom! I felt like I had gotten hit over the head with a trashcan, and I was very, very deeply upset. The story I told Brian Greene for his television program was correct: I went home; I was very nervous and very upset. My wife had tranquilizers around the house for some reason, and she said, "Take one of these and go to sleep." So I took one and I went to sleep, and then I woke up and a couple of friends came over and we had a couple of drinks, and this did not mix. I not only got drunk but I passed out, and one of my physicist friends had to pick me up off the floor and take me to bed. That was tough. It was not a nice experience.

Of course I wasn't going to leave it at rest that way; I sent it back to them and said, "Get another referee." They sent it back to me and said, "We don't get more referees." I sent it back saying, "You have to get more referees. This is important." They sent it back, saying "No we don't," and finally I sent it to another journal, which accepted it instantly. It was *Physical Review*, which is different from *Physical Review Letters*.

The discovery of string theory is usually credited to myself and Nambu. There was another version of it that was a little bit different, but the guy had the right idea, although it was a little bit less developed. His name was Holger Nielsen. He was a Dane at the Niels Bohr Institute, and he was very familiar with these kinds of ideas. A little bit later, he sent me a letter explaining his view of how it all worked, and it was a very similar idea.

After the paper came out, it was not accepted. People are very conservative about thinking pictorially like that, building models of things. They just wanted equations. They didn't like the idea that there was a physical system that you could picture behind the whole thing. It was a little bit alien to the way people were thinking at that time. This was five years before the standard model came along in '74 or '75.

The first thing that happened is that I immediately realized that this could not be a theory of hadrons. I understood why, but I also knew that the mathematics of it was too extraordinary not to mean something. It did turn out that it was not exactly the right theory of hadrons, although it's very closely related to the right theory. The idea was around for two or three years, during which it was thought that it was the theory of hadrons, exactly in that form. I knew better, but I wasn't about to go tell people, because I had my fish to fry, and I was thinking about things. I was not taken seriously at all. I was a real outsider, not embraced by the community at all.

I'll tell you the story about how I first got some credit for these things.

The already legendary Murray Gell-Mann gave a talk in Coral Gables at a big conference and I was there. His talk had nothing to do with these things. After his talk, we both went back to the motel, which had several stories to it. We got on the elevator, and sure enough the elevator got stuck with only me and Murray on it.

Murray says to me, "What do you do?"

So I said, "I'm working on this theory that hadrons are like rubber bands, these one-dimensional stringy things."

And he starts to laugh . . . and laugh. And I start to feel like, well, as my grandmother used to say, *Poopwasser.* I was so crushed by the great man's comments that I couldn't continue the conversation, so I said, "What are you working on, Murray?" And of course he said, "Didn't you hear my lecture?" Fortunately at that point the elevator started to go.

I didn't see Murray again for two years. Then there was a big conference at Fermilab, and a thousand people were there. And me, I'm still a relative nobody. And Murray is in constant competition with his colleague Richard Feynman over who is the world's greatest physicist.

As I'm standing there talking to a group of my friends, Murray walks by and in an instant turns my career and my life around. He interrupts the conversation, and in front of all my friends and closest colleagues, says, "I want to apologize to you."

I didn't know he remembered me, so I said, "What for?" He said, "For laughing at you in the elevator that time. The stuff you're doing is the greatest stuff in the world. It's just absolutely fantastic, and in my concluding talk at the conference I'm going to talk about nothing but your stuff. We've got to sit down during the conference and talk about it. You've got to explain it to me carefully, so that I get it right."

Something unimaginable had just happened to me, and I was suddenly on a cloud. So for the next three or four days at the conference, I trailed Murray around, and I would say, "Now, Murray?" And Murray would say, "No, I have to talk to somebody important."

At some point, there was a long line at the conference for people trying to talk to the travel agent. I was going to go to Israel and I had to change my ticket. It took about forty-five minutes to get to the front of the line, and when I'm two people from the front of the line, you can imagine what happens. Murray comes over and plucks me out of the line and says, "Now I want to talk. Let's talk now." Of course, I was not going to turn Murray down, so I say, "OK, let's talk," and he says, "I have fifteen minutes. Can you explain to me in fifteen minutes what this is all about?" I said OK, and we sat down, and for fourteen minutes we played a little game: He says to me, "Can you explain it to me in terms of quantum field theory?" And I say, "OK, I'll try. I'll explain it to you in terms of partons." Around 1968, Feynman had proposed that protons, neutrons, and hadrons were made of little point particles. He didn't know very much about them, but he could see in the data, correctly, that there were elements that made you think that a proton was made up out of little point particles. When you scatter protons off electrons, electrons come out. When you look at the rubbish that comes out, it tends to look as if you've struck a whole bunch of little tiny dots. Those he called partons. He didn't know what they were, that was just his name for them. Parts of protons.

Now, you have to understand how competitive Murray and Dick Feynman were. So Murray says to me, "Partons? Partons? Put-ons! Put-ons! You're putting me on!" And I thought, "What's going on here?" I had really said the wrong word. And finally he says, "What do these partons have?" I said, "Well, they have momentum. They have an electric charge." And he says, "Do they have SU(3)?" SU(3) was just a property of particles, like the electric charge is a property, or like their spin. Another property was their SU(3)-ness, which is a property that distinguishes proton from neutron. It's the thing that distinguished different particles which are otherwise very similar. Murray Gell-Mann and Yuval Ne'eman had discovered it in the early '60s, and it was what Murray became most famous for, and it led directly to the quark idea. I said, "Yeah, they can have SU-3," and he says, "Oh, you mean a quark!" So for fourteen minutes he had played this power game with me. He wanted me to say "quark," which was his idea, and not "partons," which was Dick's idea. Fourteen of the fifteen minutes had gone by and he lets me start talking, and I explained to him everything in one minute, and he looks at his watch and says, "Excuse me, but I have to talk to somebody important."

So I'm on a rollercoaster. I had gone up, down, up, down, and now I'm really down. I thought to myself, "Murray didn't understand a word I said. He's not interested. He's not going to spend his time in his lecture talking about my work," and then off in a corner somewhere I hear Murray holding forth to about fifteen people, and he's just spouting everything I told him and giving me all the credit I could hope for: "Susskind says this. Susskind says that. We have to listen to Susskind." And indeed, his talk at the end of the conference was all "Susskind this, Susskind that." And that was the start of my career. I owe Murray a lot. He's a man of tremendous integrity, and he cares about the truth, and he certainly has an interesting personality.

That jump start to my career happened around 1971. I was teaching at the Belfer Graduate School of Science, which was part of Yeshiva University, way uptown. It was an extraordinary place for a brief period of time, and it had some of the greatest theoretical physics in the world; it was outstanding. The place closed up. It went broke and I had to move to Stanford. When I went to Stanford I was an elementary-particle physicist. I was only interested in the mathematical structure of this thing. I became interested in elementary particles through it. Other people began to recognize that this was not the exactly right theory of hadrons, although it's closely related to the right theory.

I should go back a step. There were many things wrong with this theory—not wrong with the mathematical theory, but wrong in trying to compare it with nature, and to compare it with hadrons. Some of them were fixed up very beautifully by John Schwarz and Andre Neveu and a whole group of very mathematically minded string theorists, who concocted all kinds of new versions of it, and these new versions were incidentally the start of the process of discovering this incredible diversity. Each of the new versions was a little bit different, and it was always hoped that one of the new versions would look exactly like protons, neutrons, mesons, and so forth. It never happened. There were some fatal flaws.

The first was that the theory only made sense in a ridiculous number of dimensions—ten dimensions. That's not a good thing for people who live in four dimensions. That got fixed and turned out not to be so bad. The other problem was that when the theory was solved it included forces between particles that were like gravitational forces. This theory was not behaving like nuclear physics—like it was supposed to behave. It was behaving like Newtonian gravity. Particles were having forces between them that were not the kind of forces that hold a proton and neutron together but the kind of forces that hold the solar system together.

I lost a little bit of interest in it, because I was not interested at that time in gravity. John Schwarz and a number of others, including Joel Sherk, realized that this was a great opportunity. They said, "Don't think of it as a theory of hadrons, think of it as a theory of gravity." So out of a debacle, they turned it into a theory of gravitation instead of a theory of protons and neutrons. I wasn't interested at that point in gravity; I didn't know very much about gravity, and so I continued doing elementary-particle physics. Elementary-particle physicists at that time were not interested in gravity. They had no interest in gravity at all. There were people who were interested in gravity but they had no interest in string theory. So a small, isolated group of people—John Schwarz, Michael Green, Pierre Ramond and few others—carried the field on.

I became interested in it again because I became interested in black holes. Hawking had studied black holes, discovered that they radiate, that they have a temperature, that they glow, and that they give off light. I met Hawking and Gerard 't Hooft in the attic of Werner Erhard's house in San Francisco. Erhard was a fan of Sidney Coleman. Dick Feynman, myself, and David Finkelstein were his gurus. And of course we didn't give a damn about his silly business, but we loved his cigars, we loved his liquor, we loved the food that we got from him, and he was fun. He was very, very smart.

Hawking came and told us his ideas about black holes, and one of the things he told us was that things which fall into the black hole disappear from the universe completely and can never be returned, even in some scrambled form. Now, information is not supposed to be lost. It's a dictum of physics that information is preserved. What that means is that in principle you can always take a sufficiently precise look at things and figure out what happened in the past—infinitely accurately—by running them backwards.

Hawking was saying that when things fall into a black hole they're truly lost and you can never reconstruct what fell in. This violated a number of basic principles of quantum mechanics, and 't Hooft and I were stunned. Nobody else paid any attention, but we were both really stunned. I remember 't Hooft and myself were standing, glaring at the blackboard. We must have stood there for fifteen minutes without saying a word when Hawking told us these things. I was sure that Hawking was wrong. 't Hooft was sure that Hawking was wrong. And Hawking was absolutely sure that he was right in saying that information was lost inside black holes.

For thirteen years I thought about this—continuously, pretty much—and at the end of that thirteen years I began to suspect that string theory had in its guts a solution to this problem. And so I became interested again in string theory. I didn't remember anything about it. I had to go back and read my own papers, because I tried reading other people's papers and I couldn't understand them.

In the intervening years, powerful mathematics was brought to bear on the theory. I found it rather dry, since it was rather completely mathematics with very little of an intuitive, physical picture. The main things that happened were that, first of all, five versions of it were discovered. Tricks were discovered about how to get rid of the extra dimensions. You don't actually get rid of them, you curl them up into little dimensions. You can read all about that in Brian Greene's book, *The Elegant Universe*. That turned out to be a good thing.

John Schwarz and Michael Green, and a few other people, worked out the very difficult mathematics in great detail and demonstrated that the theory was not inconsistent in the ways that people thought it might be. When they showed that the mathematics was firm, Ed Witten got very excited, and once Ed Witten walked into it, well, he's a real mathematical powerhouse and dominated the field very strongly. Witten's written many famous papers, but one of his key papers, which may have been the most important one, was written in about 1990. He and collaborators around him worked out the beginnings of a mathematics of these Calabi-Yau manifolds, which are tiny, curled-up spaces that are very well explained in Brian Greene's book.

Ed is also a physicist, and he had a lot of interest in trying to make this into a real theory of elementary particles. He never quite succeeded but discovered a lot of beautiful mathematics about it. I found a lot of it rather dry, because it was not addressing physics questions the way I enjoy addressing them. It was just a little too mathematical for my taste. My taste leans less toward the mathematical and more toward the pictorial. I think in terms of pictures.

I wasn't really following the subject too closely at that point. I was still interested in black holes, and it wasn't until about 1993 that I began to suspect that there were ingredients in string theory that could resolve this puzzle of Hawking's. So at that point I really got into it. I started to think about the connection between string theory and black holes.

String theory was a theory of gravity. When you have gravity, you can have black holes, and so string theory had to have black holes in it, and it should have a resolution of this problem. Over a period of a couple of years, it did have a resolution. It did, in fact, turn out that Hawking was wrong. That is to say, he was wrong in a great way. When a person puts a finger on a problem of that magnitude, independently of whether they got it right or they got it wrong, they have a tremendous impact on the subject. And he has had a tremendous impact.

I developed some simplified ways of thinking about it that demonstrated that black holes did not lose information, that things don't fall into the black hole and disappear, they eventually come back out. They are all scrambled up, but nevertheless they come back out. I began writing papers on that, and my paper, which said that stuff does not get lost inside a black hole in string theory, stimulated the string community to start thinking about black holes. There was an eruption of papers—mine, Joe Polchinski's, Andy Strominger's, Cumrun Vafa's—that really nailed that problem down. And black holes have been solved. Black holes have been understood. To this day the only real physics problem that has been solved by string theory is the problem of black holes. It led to some extremely revolutionary and strange ideas.

Up to now string theory has had nothing to say about cosmology. Nobody has understood the relationship between string theory and the Big Bang, inflation, and other aspects of cosmology. I frequently go to conferences that often have string theorists and cosmologists, and usually the string theory talks consist of apologizing for the fact that they haven't got anything interesting to tell the cosmologists. This is going to change very rapidly now, because people have recognized the enormous diversity of the theory.

People have been trying to do business the old way. With string theory, they were trying to do the things that they would have done with the earlier theories and it didn't make a lot of sense for them to do so. They should have been looking at what's really unique and different about string theory, not what looks similar to the old kind of theories. And the thing which is really unique and very, very special is that it has this diversity, that it gives rise to an incredibly wild number of different kinds of environments that physics can take place in.

11
Smolin vs. Susskind:
The Anthropic Principle

INTRODUCTION by John Brockman

In the summer of 2004, I received a copy of an email sent by Leonard Susskind to a group of physicists which included an attached file entitled "Answer to Smolin." This was the opening salvo of an intense email exchange between Susskind and Smolin concerning Smolin's argument that "the Anthropic Principle (AP) cannot yield any falsifiable predictions, and therefore cannot be a part of science."

After reading several postings by both physicists, I asked them if (a) they would consider posting some of the comments on *Edge. org*, and (b) if they would each write a new, and final, "letter."

Both agreed, but only after a negotiation: (1) No more than one letter each; (2) neither sees the other's letter in advance; (3) no changes after the fact. A physics shoot-out.

While this is a conversation conducted by physicists for physicists, it should nonetheless be of interest to *Edge* readers, as it's in the context of previous *Edge* features with the authors, it's instructive as to how science is done, and it's a debate that clarifies, not obfuscates. And finally it's a good example of what *Edge* is all about, where contributors share the boundaries of their knowledge and experience with each other and respond to challenges, comments, criticisms, and insights. The constant shifting of metaphors, the intensity with which we advance our ideas to each other—this is what intellectuals do. *Edge* draws attention to the larger context of intellectual life.

On July 26, 2004, Lee Smolin published a paper (hep-th/0407213, "Scientific alternatives to the anthropic principle"). He emailed Leonard Susskind, asking for a comment. Not having had a chance to read the paper, Susskind asked Smolin if he would summarize the arguments. Here is Smolin's email summary:

Dear Lenny,

Thanks. Glad to. I'll start with one of the main arguments I make.

I show that the argument of Weinberg and others [Garriga and Vilenkin] are incorrect. The subtle point is that their arguments have embedded in them correct arguments having to do only with what we observe. To an already correct argument is then added mention of the anthropic principle. As it is added to an already correct argument, the anthropic principle plays no role in the actual scientific argument.

Here is how it goes: We start with a theory of structure formation that tells us "Too large positive Lambda interferes with galaxy formation."

We do observe that galaxies have formed. Therefore we predict that the cosmological constant could not have been too large. This is correct reasoning, and it agrees with observation.

What Weinberg and others have done is to make the error of embedding this argument into one that first mentions some version of the anthropic principle. The mistake is not to notice that because it has been added to an argument that is already correct, the mention of the anthropic principle [or the principle of mediocrity, or life] plays absolutely no role in the argument.

The logic of their arguments is: A implies B; B is observed; B, together with theory C, implies D.

Here A is any form of the anthropic principle or principle of mediocrity, together with assumptions about priors, probability distributions on universes etc., plus our own existence, that leads to the conclusion that we should observe B.

B is that galaxies have formed.

C is the theory of structure formation.

D is that the cosmological constant is not too large.

The fallacy is not to recognize that the first line plays no role in the argument, and the prediction of D is equally strong if it is dropped. One can prove this by noting that if D were not seen, one would have to question the theory C [assuming the observation is correct, as it certainly is here]. One would have no reason to question either A or the assertion that A implies B.

This is the same fallacy involved in Hoyle's argument about carbon. He reasoned simply from an observation that carbon is plentiful in our universe to a prediction that, as it must have been formed in stars, there must be a resonance at a particular energy. This was correct and the resonance was observed. But he fallaciously attributed the argument to the existence of life, which was a non sequitur.

In the paper, I show that every use of the anthropic principle claimed in physics and cosmology is either an example of this fallacy or is so vague that one can get any conclusion one wants, and match any observation, by manipulating the assumptions made.

I then go on to explain what a multiverse theory would have to do to yield genuine falsifiable predictions that actually depend on the existence of the multiverse. I give conditions for this to work. I then show that there exist real theories that satisfy these conditions, one of them being my old natural selection idea.

Therefore, the anthropic principle cannot help us to do science. But there are ways to do science if we are faced with a multiverse.

Lee

July 29, 2004: Susskind paper on Smolin's theory of cosmic natural selection: Cosmic Natural Selection hep-th/0407266

L. Susskind
Department of Physics
Stanford University
Stanford, CA 94305-4060

Abstract:
I make a number of comments about Smolin's theory of Cosmic Natural Selection.

In an unpublished note [the "Answer to Smolin" email] I criticized Smolin's theory of cosmological natural selection [1] which argues that we live in the fittest of all possible universes. By fitness, Smolin means the ability to reproduce. In my criticism I used the example of eternal inflation which is an extremely efficient reproduction mechanism. If Smolin's logic is applied to that example it would lead to the prediction that we live in the universe with the maximum cosmological constant. This is clearly not so.

Smolin proposes that the true mechanism for reproduction is a bouncing black hole singularity that leads to a new universe behind the horizon of every black hole. Thus Smolin suggests that the laws of nature are determined by maximizing the number of black holes in a universe.

Smolin also argues that it is not obviously wrong that our physical parameters, including the smallness of the cosmological constant, maximize the black hole formation. To make sense of this idea, one must assume that there is a very dense discretuum of possibilities, in other words a rich landscape of the kind that string theory suggests [4][5][6][7].

The detailed astrophysics that goes into Smolin's estimates is extremely complicated—too complicated for me—but the basic theoretical assumptions that go into the theory can be evaluated, especially in light of what string theory has taught us about the landscape and about black holes.

As I said, there are two mechanisms, eternal inflation and black hole production that can contribute to reproduction, and it is important for Smolin's scenario that black holes dominate. Considering the low density of black holes in our universe and the incredible efficiency of exponential inflation, it seems very hard to believe that black holes win unless eternal inflation is not possible for some reason.

Smolin argues that we know almost nothing about eternal inflation but we know a great deal about black holes including the fact that they really exist. This is a bit disingenuous. Despite a great deal of serious effort [8] [9], the thing we understand least is the resolution of black hole and cosmic singularities. By contrast, eternal inflation in a false vacuum is based only on classical gravity and semiclassical Coleman–De Luccia bubble nucleation [2][3].

The issue here is not whether the usual phenomenological inflation was of the eternal kind although that is relevant. Eternal inflation taking place in any false vacuum minimum on the landscape would favor [in Smolin's sense] the maximum cosmological constant. But for the sake of argument I will agree to ignore eternal inflation as a reproduction mechanism.

Lee Smolin and Leonard Susskind

The question of how many black holes are formed is somewhat ambiguous. What if two black holes coalesce to form a single one? Does that count as one black hole or two? Strictly speaking, given that black holes are defined by the global geometry, it is only one black hole. What happens if all the stars in the galaxy eventually fall into the central black hole? That severely diminishes the counting. So we better assume that the bigger the black hole, the more babies it will have. Perhaps one huge black hole spawns more offspring that 10^{22} stellar black holes.

That raises the question of what exactly is a black hole? One of the deepest lessons we have learned over the past decade is that there is no fundamental difference between elementary particles and black holes. As repeatedly emphasized by 't Hooft [10][11] [12], black holes are the natural extension of the elementary particle spectrum. This is especially clear in string theory where black holes are simply highly excited string states. Does that mean that we should count every particle as a black hole?

Smolin's theory requires not only that black hole singularities bounce but that the parameters such as the cosmological constant suffer only very small changes at the bounce. This I find not credible for a number of reasons. The discretuum of string theory does indeed allow a very dense spectrum of cosmological constants but neighboring vacua on the landscape do not generally have close values of the vacuum energy. A valley is typically surrounded by high mountains, and neighboring valleys are not expected to have similar energies.

Next—the energy density at the bounce is presumably Planckian. Supposing that a bounce makes sense, the new universe starts with Planckian energy density. On the other hand Smolin wants the final value of the vacuum energy density to be very close to the original. It sounds to me like rolling a bowling ball up to the top of a very high mountain and expecting it to roll down, not to the original valley, but to one out of 10^{120} with almost identical energy. I find that unlikely.

Finally, we have learned some things about black holes over the last decade that even Stephen Hawking agrees with [13]. Black holes do not lose information. The implication [14] is that if there is any kind of universe creation in the interior of the black hole, the quantum state of the offspring is completely unique and can have no memory of the initial state. That would preclude the kind of slow mutation rate envisioned by Smolin.

Smolin seems to think that there is significant evidence that singularity resolution [by bounce] is imminent. Loop quantum gravity, according to him, is on the threshold of accomplishing this. Perhaps it will. But either it will be consistent with information conservation in which case the baby can have no memory of the parent, or it will not. If not, it probably means that loop gravity is inconsistent.

A. Guth and L. Susskind, To be published.

References:

[1] Lee Smolin, Scientific alternatives to the anthropic principle, hep-th/0407213.

[2] S. R. Coleman and F. De Luccia, *Phys. Rev. D* 21, 3305 [1980].

[3] S. K. Blau, E. I. Guendelman. and A. H. Guth, "The Dynamics Of False Vacuum Bubbles," *Phys. Rev. D* 35, 1747 [1987].

[4] Raphael Bousso, Joseph Polchinski, Quantization of four-form fluxes and dynamical neutralization of the cosmological constant, hep-th/0004134, JHEP 0006 [2000] 006.

[5] Shamit Kachru, Renata Kallosh, Andrei Linde, Sandip P. Trivedi, De Sitter vacua in string theory, hep-th/0301240.

[6] Leonard Susskind, The anthropic landscape of string theory, hep-th/0302219.

[7] Michael R. Douglas, The statistics of string/M vacua, hep-th/0303194, JHEP 0305 [2003] 046; Sujay Ashok, Michael R. Douglas, Counting flux vacua, hep-th/0307049; Michael R. Douglas, Bernard Shiffman, Steve Zelditch, Critical points and super-symmetric vacua math. CV/0402326.

[8] G. T. Horowitz and J. Polchinski, *Phys. Rev. D* 66, 103512 [2002] [arXiv:hep-th/0206228].

[9] L. Fidkowski, V. Hubeny, M. Kleban, and S. Shenker, The black hole singularity in AdS/CFT, JHEP 0402, 014 [2004] [arXiv:hep-th/0306170].

[10] G. 't Hooft, "The unification of black holes with ordinary matter." Prepared for Les Houches Summer School on Gravitation and Quantizations, Session 57, Les Houches, France, 5 Jul – 1 Aug 1992.

[11] L. Susskind, Some speculations about black hole entropy in string theory, arXiv:hep-th/9309145.

[12] G. T. Horowitz and J. Polchinski, "A correspondence principle for black holes and strings," *Phys. Rev. D* 55, 6189 [1997] [arXiv:hep-th/9612146].

[13] *New York Times* 7/22/04.

[14] G. T. Horowitz and J. Maldacena, The black hole final state, JHEP 0402, 008 [2004] [arXiv:hep-th/0310281].

July 29, 2004, Smolin's email response to Susskind's criticisms:

Dear Lenny and colleagues,

I am grateful to Lenny for taking the time to respond to my paper. I will be as brief as I can in replying, especially as the key points are already presented in detail in my paper hep-th/0407213 ["Scientific alternatives to the anthropic principle"] or in my book, *The Life of the Cosmos*, or previous papers on the subject.

For clarity, I had in section 5.1.6 identified two arguments in Weinberg's papers. The first is the one I criticized in the summary. Susskind reponds, reasonably, by agreeing, and then raising the second argument. This argument is also criticized in detail in my paper, and it was perhaps a mistake not to include this in the summary I sent to Susskind.

This second argument is based on a version of the AP called the principle of mediocrity by Garriga and Vilenkin, who have done the most to develop it. Their version states that " . . . our civilization is typical in the ensemble of all civilizations in the universe."

This argument is discussed in full in sections 5.1.5 and 5.1.6. There I argue that the mediocrity principle cannot yield falsifiable predictions because it depends on the definition of the ensemble within which our civilization is taken to be typical, as well as on assumptions about the probability distribution. I establish this by general argument as well as by reference to specific examples, including Weinberg's use of it.

Can this be right if, as Susskind claims, Weinberg's prediction was found to hold? In fact, Weinberg's prediction did not work all that well. In the form that he made it, it led to an expectation of a cosmological constant larger than the observed value. Depending on the ensemble chosen and the assumptions made about the probability distribution, the probability that Lambda be as small as observed ranges between about 10 percent and a few parts in ten thousand. In fact, the less probable values are the more reasonable, as they come from an ensemble where Q, the scale of the density fluctuations, is allowed to vary. While I am not an

Lee Smolin and Leonard Susskind

expert here, it appears from a reading of the literature [references in the paper] that to make the probability for the present value as large as 10 percent one has to assume that Q is frozen and fixed by fundamental theory. It is hard to imagine a theory where the parameters vary but Q does not, as it depends on parameters in the inflaton potential.

But, because there is so much flexibility—and an absence of strict, up or down, falsifiable predictions—anyone who wants to continue to use the AP in this context is free to modify the assumptions about the prior probability distribution to raise the probability for the observed value of the vacuum energy from 10^{-4} to order unity. No one can prove they are wrong to do so—and this is precisely the problem.

I do believe it is important to insist on falsifiability, because it alone prevents we theorists from keeping theories alive indefinitely, by freely adjusting them to match data.

It was worry about the possibility that string theory would lead to the present situation, which Susskind has so ably described in his recent papers, that led me to invent the Cosmological Natural Selection [CNS] idea and to write my first book. My motive, then as now, is to prevent a split in the community of theoretical physicists in which different groups of smart people believe different things, with no recourse to come to consensus by rational argument from the evidence.

The CNS idea was invented, not for itself, but to give an existence proof that shows that the anthropic principle can be replaced by a falsifiable theory, that explains everything the AP claims to. The reason I chose the term "landscape" of string theories, is to anticipate the transition to "fitness landscapes," a term that comes from mathematical models that explain why the mechanism of natural selection is falsifiable.

As the theory of CNS is falsifiable, it is vulnerable to criticisms of the kind Susskind makes. Let me briefly address them.

The last first. Susskind claims that life is exceptional in the ensemble of universes. This is not true in CNS. The whole point of cosmological natural selection is that it follows the logical schema described in section 5.1.4 and 5.2. There, and in more detail in the book and papers, I show that falsifiable predictions can be gotten from a multiverse theory if the distribution of universes is very different from random. CNS results in a distribution peaked around small regions of the parameter space—so that a typical universe in this distribution is very untypical in any randomly chosen ensemble. I show in detail why falsifiability follows from this. I also show why reproduction through black holes leads to a multiverse in which the conditions for life are common— essentially because some of the conditions life requires, such as plentiful carbon, also boost the formation of stars massive enough to become black holes.

Next, there is a big difference between the explanatory power of the reproduction mechanisms of eternal inflation and black holes bouncing. This stems from the fact that any selection mechanism can only operate to tune parameters that strongly affect the rate of reproduction. Given that standard inflation acts on grand unified scales, the differential reproduction rate due to eternal inflation is only sensitive to the parameters that govern GUT [Grand Unified Theory]–scale physics plus the vacuum energy. Thus, eternal inflation cannot explain the values of any of the low-energy parameters, such as the masses of the light quarks and leptons. This means it cannot explain why there are long-lived stars or many stable nuclear bound states leading to a complex chemistry.

Reproduction through black holes explains all the puzzles and coincidences of low-energy physics because carbon chemistry,

Lee Smolin and Leonard Susskind

long-lived stars, etc., are essential for the mechanisms that lead to the formation of massive stars—those that become black holes. There is a long list of observed facts this turns out to explain, and a few genuine predictions. These are summarized in the paper and discussed at length in the book.

With regard to cosmological natural selection and the cosmological constant: If both mechanisms of reproduction exist, there is a competition between them that determines Lambda. Eternal inflation favors a larger cosmological constant, as Lenny says. But by Weinberg's first argument, if Lambda is too big, there are no galaxies, hence many fewer black holes. To my knowledge, nobody has attempted to do a detailed analysis including both mechanisms. Vilenkin has pointed out that too-small Lambda can hurt black-hole production in a single universe by making mergers of spirals more common. This is briefly discussed in section 6.2, where I propose that the observed value may maximize the production of black holes—but this has not been analyzed in any detail.

Certainly, if the only mechanism of reproduction is eternal inflation, cosmological natural selection is wrong. But we know that there are black holes, and we have reasonable theoretical evidence that black-hole singularities bounce. I expect that in the next year we may have reliable quantum gravity calculations that will settle the issue, building on the methods Bojowald and collaborators have used to study cosmological bounces. So the consequences of reproduction through black holes seem reasonable to explore. Not only that, we are on firm ground when we do so, because star and black-hole formation are observed and controlled by known physics and chemistry.

We know much less about eternal inflation. We cannot observe whether it takes place or not, and there is little near-term chance to check the theories that lead to it independently, as there are alternative early universe theories—inflationary and not—that agree with all the cosmological data and do not yield reproduction through eternal inflation.

I do not know if CNS is the only way to get a falsifiable multiverse theory; it is just the only way I've been able to think of. As I have been saying for more than ten years, if CNS can be proved wrong, and if people are stimulated to invent more falsifiable theories to explain the observed parameters, this would be all to the good. So far CNS has not been falsified, but I read astro-ph every day looking for the discovery of a very massive neutron star that will disprove it.

I am very glad that Susskind has been able to give these issues much more visibility. But it would be very unfortunate if string theorists finally accept that there is an issue with predictability, only to fall for the easy temptation of adopting a strategy towards it that cannot yield falsifiable theories. The problem with nonfalsifiable theories is nothing other than that they cannot be proven wrong. If a large body of our colleagues feels comfortable believing a theory that cannot be proved wrong, then the progress of science could get stuck, leading to a situation in which false but unfalsifiable theories dominate the attention of our field.

Thanks,

Lee

Leonard Susskind's "Final Letter"

When I was asked if I would be willing to continue a debate with Lee Smolin on the *Edge* website, my initial reaction was to say no. The problem is that the easiest ideas to explain, which sound convincing to a general audience, are not always the best ideas. The unwary layman says to himself, "Yeah, I understand that. Why is this other guy making it so complicated?" Well, the answer is that those simple ideas, that sound like you understand them, often have deep technical flaws, and the correct ideas can be very difficult to explain. All a person like myself can do is to say, "Trust me. I know what I'm doing, and he doesn't. And besides, so-and-so agrees with me." That doesn't make a good impression. It can be a no-win situation.

Why did I agree to do it? Partly because I love explaining physics. Mostly—I don't know why. But here goes nothing as they say.

In a nutshell, here is the view of physics and cosmology that Smolin is attacking:

(1.) In the remote past, the universe inflated to an enormous size, many orders of magnitude bigger than the observed portion that we can see. Most of the universe is behind the cosmic horizon and cannot be directly detected.

(2.) The mechanism of inflation leads to a diverse universe filled with what Alan Guth calls pocket universes (PU's). We live in one such PU. Some people call this super-universe the "Multiverse." I like the term "Megaverse." This growth and continuous spawning of pocket U's is called, in the trade, eternal inflation.

(3.) String theory leads to a stupendously large "Landscape" of possibilities for the local laws of nature in a given pocket. I will call these possibilities "environments." Most environments are very different from our own and would not permit life—at least, life as we know it.

(4.) Combining 1, 2, and 3—the universe is a megaverse filled with a tremendously large number of local environments. Most of the volume of the megaverse is absolutely lethal to life. Some small fraction is more hospitable. We live somewhere in that fraction.

That's it.

There are good reasons for believing 1 − 4, based on a combination of theoretical and experimental physics. In fact, I don't know anyone who disagrees with 1. Assumption 2 is not quite a consequence of 1, but it's difficult to avoid 2 in conventional inflation theories.

The physics that goes into it is a very familiar application of trustworthy methods in quantum field theory and general relativity. It's called Coleman–De Luccia semi-classical tunneling by instantons, based on a famous paper by the incomparable Sidney Coleman and his collaborator Frank De Luccia. It is the same physics that has been used from the 1930s to explain the decay of radioactive nuclei.

String theorists are split on whether 3 is a good thing or a bad thing but not about whether it is correct. Only one string theorist seriously challenged the technical arguments, and he was wrong. In any case, Smolin and I agree about 3. I think we also agree that most of the Landscape is totally lethal to life, at least life of our kind. Finally 4. There's the rub. As far as I am concerned, 4 is simply 1+2+3. But Smolin has other ideas and 4 just gets in the way.

Let's suppose for the moment that these four points are correct. What then determines our own environment? In other words, why do we find ourselves in one kind of PU rather than another? To get an idea of what the issues are in a more familiar context, let's replace 1 - 4 with analogous points regarding the ordinary known universe.

(1'.) The universe is big—about 15 billion light-years in radius.

(2'.) The expansion of the universe led to a huge number of

condensed astronomical objects—at minimum, 10^{23} solar sys-. tems.

(3'.) The laws of gravity, nuclear physics, atomic physics, chemistry, and thermodynamics allow a very diverse set of possible environments, from the frozen cold of interstellar space to the ferocious heat of stellar interiors, with planets, moons, asteroids, and comets somewhere in between. Even among planets, the diversity is huge—from Mercury to Pluto.

(4'.) The universe is filled with these diverse environments, most of which are lethal. But the universe is so big that, statistically speaking, it is very likely that one or more habitable planets exists.

I don't think anyone questions these points. But what is it that decides which kind of environment we live in—the temperature, chemistry, and so on? In particular what determines the fact that the temperature of our planet is between freezing and boiling? The answer is that nothing does. There are environments with temperatures ranging from almost absolute zero to trillions of degrees. Nothing determines the nature of our environment— except for the fact that we are here to ask the question! The temperature is between freezing and boiling because life (at least our kind) requires liquid water. That's it. That's all. There is no other explanation. [1]

This rather pedestrian, commonsense logic is sometimes called the anthropic principle. Note that I mean something relatively modest by the AP. I certainly don't mean that everything about the laws of physics can be determined from the condition that life exists—just those things that turn out to be features of the local environment and are needed to support life.

Let's imagine that the Earth was totally cloud-bound or that we lived at the bottom of the sea. Some philosopher who didn't like these ideas might object that our hypotheses 1' - 4' are unfalsifiable. He might say that since there is no way to observe these other regions with their hostile environments—not without penetrating the impenetrable veil of clouds—the theory is unfalsifiable. That, according to him, is the worst sin a scientist can commit. He will say, "Science means falsifiability. If a hypothesis can't be proved false it is not science." He might even quote Karl Popper as an authority.

From our perspective, we would probably laugh at the poor deluded fellow. The correctness of the idea is obvious and who cares if they can falsify it.

Even worse, he wouldn't even be correct about the falsifiability. Here is a way that the anthropic reasoning might be proved false without penetrating the veil of clouds: Suppose an incredibly accurate measurement of the average temperature of the Earth gave the answer (in centigrade)T=50.00000000000000000000 00000000000000000000000000000000000000000000000000 00000000000000000000000000000000 degrees. In other words, the temperature was found to be exactly midway between freezing and boiling, to an accuracy of one hundred decimal places. I think we would be justified in thinking that there's something beyond the anthropic principle at work. There is no reason, based on the existence of life, for the temperature to be so symmetrically located between boiling and freezing. So discovering such a temperature would pretty convincingly mean that the existence of life is not the real reason why the temperature is between 0 and 100 degrees.

Smolin's chief criticism of 1 - 4 is that they are unfalsifiable. But it isn't hard to think of ways of falsifying the anthropic principle. In particular, Weinberg's prediction that if the anthropic principle is true, then the cosmological constant should not be

Lee Smolin and Leonard Susskind

exactly zero, is very similar to the example I just invented. Weinberg attempted to falsify the anthropic principle. He failed. The anthropic principle survived. You can read about the details in Weinberg's book *Dreams of a Final Theory*.

By "unfalsifiable," Smolin probably means that other pocket universes can never be directly observed because they are behind an impenetrable veil—i.e., the cosmic event horizon. Throughout my long experience as a scientist, I have heard unfalsifiability hurled at so many important ideas that I am inclined to think that no idea can have great merit unless it has drawn this criticism. I'll give some examples.

From psychology: You would think that everybody would agree that humans have a hidden emotional life. B. F. Skinner didn't. He was the guru of a scientific movement called behaviorism that dismissed anything that couldn't be directly observed as unscientific. The only valid subject for psychology, according to the behaviorist, is external behavior. Statements about the emotions or the state of mind of a patient were dismissed as unfalsifiable. Most of us today would say that this is a foolish extreme.

From physics: In the early days of the quark theory, its many opponents dismissed it as unfalsifiable. Quarks are permanently bound together into protons, neutrons, and mesons. They can never be separated and examined individually. They are, so to speak, hidden behind a different kind of veil. Most of the physicists who made these claims had their own agendas, and quarks just didn't fit in. But by now, although no single quark has ever been seen in isolation, there is no one who seriously questions the correctness of the quark theory. It's part of the bedrock foundation of modern physics.

Another example is Alan Guth's inflationary theory. In 1980 it seemed impossible to look back to the inflationary era and see direct evidence for the phenomenon. Another impenetrable veil called the "surface of last scattering" prevented any observation of the inflationary process. A lot of us did worry that there might be no good way to test inflation. Some—usually people with competing ideas—claimed that inflation was unfalsifiable and therefore not scientific.

I can imagine the partisans of Lamarck criticizing Darwin, "Your theory is unfalsifiable, Charles. You can't go backward in time, through the millions of years over which natural selection acted. All you will ever have is circumstantial evidence and an unfalsifiable hypothesis. By contrast, our Lamarckian theory is scientific because it is falsifiable. All we have to do is create a population that lifts weights in the gym every day for a few hours. After a few generations, their children's muscles will bulge at birth." The Lamarckists were right. The theory is easily falsified—too easily. But that didn't make it better than Darwin's theory.

There are people who argue that the world was created 6,000 years ago, with all the geological formations, isotope abundances, dinosaur bones, in place. Almost all scientists will point the accusing finger and say, "Not falsifiable!" I'm sure that Smolin would agree with them, and so would I. But so is the opposite—that the universe was not created this way—unfalsifiable. In fact, that is exactly what creationists do say. By the rigid criterion of falsifiability, "creation-science" and science-science are equally unscientific. The absurdity of this position will, I hope, not be lost on the reader.

Good scientific methodology is not an abstract set of rules dictated by philosophers. It is conditioned by, and determined by, the science itself and the scientists who create the science. What may have constituted scientific proof for a particle physicist of

Lee Smolin and Leonard Susskind

the 1960s—namely, the detection of an isolated particle—is inappropriate for a modern quark physicist, who can never hope to remove and isolate a quark. Let's not put the cart before the horse. Science is the horse that pulls the cart of philosophy.

In each case that I described—quarks, inflation, Darwinian evolution—the accusers were making the mistake of underestimating human ingenuity. It took only a few years to indirectly test the quark theory with great precision. It took 20 years to do the experiments that confirmed inflation. And it took 100 years or more to decisively test Darwin (some would even say that it has yet to be tested). The powerful methods that biologists would discover a century later were unimaginable to Darwin and his contemporaries. What people usually mean when they make the accusation of unfalsifiability is that they themselves don't have the imagination to figure out how to test the idea. Will it be possible to test eternal inflation and the Landscape? I certainly think so, although it may be, as in the case of quarks, that the tests will be less direct, and involve more theory, than some would like.

Finally, I would point out that the accusation of unfalsifiability is being thrown by someone with his own agenda. Smolin has his own theory, based on ideas about the interior of black holes. There is, of course, absolutely nothing wrong with that, and Smolin is completely candid about it.

Smolin believes (as I and most cosmologists do) that there is a sense in which the universe, or perhaps I should say "a universe," can reproduce, parent universes spawning baby universes. Perhaps here is a good time to talk about a linguistic point. The word "universe" was obviously intended to refer to all that exists. It was not a word that was intended to have a plural. But by now, physicists and cosmologists have gotten used to the linguistic discord. Sometimes we mean all that exists, but sometimes we mean an expanding region of space with particular properties. For example, we might say that in our universe the electron is lighter than the proton. In some other distant universe, the electron is heavier than the proton. Guth's term—pocket universe—may be a better term but it tends to ruin the prose.

Although we agree that some form of universe-reproduction can occur, Smolin and I disagree about the mechanism. Just the ordinary expansion of the universe is a form of reproduction. For example, if the radius of the universe doubles, you can either picture each cubic meter stretching to 8 cubic meters or you can say that the original cubic meter gave birth to seven children. Inflation is the exponential expansion of space. It can be understood as an exponentially increasing population of regions. Moreover, according to absolutely standard principles, some of the offspring can be environmentally different than the parent. In this sense, a population of PU's exponentially reproduces as the universe inflates. The modern idea of eternal inflation is that the universe eternally inflates, endlessly spawning PU's, such as our own. The analogy with the tree of life is apt. Any species eventually becomes extinct, but the tree keeps on growing by shooting off new branches and twigs. In the same way, a given PU will eventually end, but eternal inflation goes on. Just as the population of organisms will be numerically dominated by the fastest reproducers (bacteria), the volume of space will be dominated by the most rapidly inflating environment: an environment, I might add, that is totally lethal. If eternal inflation is part of the story of the universe, we can conclude that our local environment is by no means typical. The typical region of space will be one with the largest possible cosmological constant. For Smolin's alternative reproduction mechanism to be relevant, eternal inflation must not occur, for some unknown reason.

Lee Smolin and Leonard Susskind

Smolin's picture of reproduction is that it takes place in the interior of black holes. He believes that in the deep interior of every black hole, the dreaded singularity is a source of a new universe that arises out of the infinitely compressed and heated matter as it contracts and (according to Smolin) subsequently rebounds and expands. By this hypothetical mechanism, a new infant universe is created inside the black hole. The idea is that the child can expand and grow into a genuine adult universe, all hidden from the parent, behind the horizon. Moreover the child universe must have different properties than the parent for Smolin's cosmological natural selection to work. Random mutation is a necessary ingredient to natural selection.

Smolin adds one more assumption that follows the biological paradigm. In ordinary biology, the child inherits information about the parents' traits through the genetic code, which may be altered by mutation but only a tiny bit. Smolin must assume that the offspring only differ by very small amounts from the parent. More precisely, he assumes that the constants of physics in the offspring universe are almost the same as in the parent universe. Without this assumption natural selection wouldn't work.

And what does this setup select for? As in life, evolution selects for maximal ability to reproduce. This, according to Smolin, means that PU's whose properties maximize the tendency to produce black holes will dominate the population. So Smolin argues that our laws and constants of nature are tuned to values that maximize black-hole production. According to Smolin, no anthropic reasoning is needed, and that makes his theory "scientific."

This is an extremely clever idea. You can read about it in one of Smolin's papers that you can find on the net. Open your web browser to http://www.arxiv.org. That's where physicists publish their work these days. On the General Relativity and Quantum

Cosmology archive, look up gr-qc/9404011. That's one of the first papers that Smolin wrote on "Cosmological Natural Selection." The paper, from 1994, is clear and enjoyable to read. But for some reason it hasn't caught on with either physicists or cosmologists. In fact, when I went to track down subsequent papers on the subject to see if new developments had taken place, I found that there were only 11 citations to the paper. Four of them were by Smolin and two others were critical of the idea—one incorrectly so, in my judgment.

I'm not sure why Smolin's idea didn't attract much attention. I actually think it deserved far more than it got. But I do know why I was skeptical. Two details, one very technical and one not so technical, seem to me to undermine the idea.

The first, not so technical objection: Frankly, I very much doubt that our laws maximize the number of black holes in the universe. In fact, the meaning of the number of black holes in a given universe is unclear. Suppose that in our pocket universe every star collapses to a black hole eventually. Then in the part of the universe we presently observe, the number of black holes will eventually be about 10^{22}. But suppose that as time goes on, all the black holes in a given galaxy eventually fall into a central black hole at the galactic center. Then the final number will be more like the number of galaxies—about 10^{11}. Smolin of course prefers the larger number, since he wants to argue that our universe has more black holes than any other possible universe. But strictly speaking, according to the rigorous definition of a black hole, the smaller number is the correct one. But let me be generous and use a looser definition of "black hole," so that anything that temporarily looked like a black hole is counted. But with this rule, it is easy to change the laws of physics so that many more black holes would have been present in the past.

Lee Smolin and Leonard Susskind

If, for example, the minute density contrasts in the early universe, which had the unnaturally small numerical value of 10^{-5}, were not so weak, the universe would have been dominated by small black holes. Those black holes might have coalesced into larger black holes, but I said I would be generous and count them all.

Combine the increase of density contrast with an increase in the strength of gravity and a rapid inflation prehistory and you can make stupendous numbers of black holes. In fact, if gravity were made as strong as it could reasonably be, every elementary particle except photons and gravitons would be a black hole!

I have exactly the opposite opinion from Smolin's. If the universe were dominated by black holes, all matter would be sucked in and life would be completely impossible. It seems clear to me that we live in a surprisingly smooth world remarkably free of the ravenous monsters that would devour life. I take the lack of black holes to be a sign of some anthropic selection.

Now I come to one of those technical objections, which I think is quite damning but which may mean very little to a layman. Smolin's idea is tied to Hawking's old claim that information can fall into a black hole and get trapped behind the horizon. Smolin requires a great deal of information to be transferred from the parent universe to the infant at the bouncing singularity. But the last decade of black-hole physics and string theory have told us that NO information can be transferred in this way!

Some readers may recognize the issue I'm talking about. Anyone who has read the recent *New York Times* article by Dennis Overbye knows that the ultimate fate of information falling into a black hole was the subject of an long debate involving Stephen Hawking, myself, the famous Dutch physicist Gerard 't Hooft and many other well known physicists. Hawking believed that information does disappear behind the horizon, perhaps into a baby universe. This would be consistent with Smolin's idea that offspring universes, inside the black hole, remember at least some of the details of the mother universe. My own view and 't Hooft's was that nothing can be lost from the outside world—not a single bit. Curiously the cosmological debate about cosmological natural selection revolves around the same issues that came to the attention of the press a week or two ago. The occasion for the press coverage was Hawking's recantation. He has reversed his position.

Over the last decade, since Smolin put forward his clever idea, the black-hole controversy has largely been resolved. The consensus is that black holes do not lose any information. I'll cite some of the most influential papers that you can look up yourself: hep-th 9309145, hep-th 9306069, hep-th 9409089, hep-th 9610043, hep-th 9805114, hep-th 9711200. Incidentally, the combined total number of citations for these six papers is close to 6,000. Another paper, coauthored very recently by the author of one of these classics, directly attacks Smolin's assumption. In fact it was one of the 11 papers I found citing Smolin's paper. If you want to look it up, here is the archive reference: hep-th 0310281. I warned you that I would say "And besides, so-and-so agrees with me." I apologize, but at least you can go check for yourself.

The implication of these papers is that no information about the parent can survive the infinitely violent singularity at the center of a black hole. If such a thing as a baby universe makes any sense at all, the baby will have no special resemblance to the mother. Given that, the idea of an evolutionary history that led by natural selection to our universe makes no sense.

I'm sure there are physicists who are unconvinced by the arguments of the abovementioned papers, despite the number of citations. They have all the right in the world to be skeptical, but the average reader of this page should know that these people are swimming against the tide.

Finally let me quote a remark of Smolin's that I find revealing. He says, "It was worry about the possibility that string theory would lead to the present situation, which Susskind has so ably described in his recent papers, that led me to invent the Cosmological Natural Selection (CNS) idea and to write my first book. My motive, then as now, is to prevent a split in the community of theoretical physicists in which different groups of smart people believe different things, with no recourse to come to consensus by rational argument from the evidence."

First of all, preventing a "split in the community of theoretical physicists" is an absurdly ridiculous reason for putting forward a scientific hypothesis.

But what I find especially mystifying is Smolin's tendency to set himself up as an arbiter of good and bad science. Among the people who feel that the anthropic principle deserves to be taken seriously are some famous physicists and cosmologists with extraordinary histories of scientific accomplishment. They include Steven Weinberg [2], Joseph Polchinski [3], Andrei Linde [4], and Sir Martin Rees [5]. These people are not fools, nor do they need to be told what constitutes good science.

References:
[1] Of course, you might say that the distance to the sun determines the temperature. But that just replaces the question with another, "Why is our planet at the precise distance that it is?"
[2] Professor of physics, University of Texas, and 1979 Nobel Prize winner.
[3] Professor of physics, Kavli Institute for Theoretical Physics.

[4] Professor of physics, Stanford University; winner of many awards and prizes, including the Dirac Medal and Franklin Medal.

[5] Astronomer Royal of Great Britain.

Lee Smolin's "Final Letter"

I am very pleased that Lenny Susskind has taken the time to respond to my paper on the anthropic principle (AP) ["Scientific alternatives to the anthropic principle"] and to discuss cosmological natural selection (CNS). Susskind is for me the most inspiring figure of his generation of elementary-particle physicists. Indeed, the initial ideas that became loop quantum gravity came from applying to quantum gravity some of what I had learned from his work on gauge theories. And when in the late 1990s I began to work again on string theory, it was because of papers of his describing how special relativity was compatible with string theory.

I was thus extremely pleased when Susskind began arguing for a view of string theory I came to some time ago—that there is not one theory but a "landscape" of many theories. But I was equally disturbed when he and other string theorists embraced versions of the anthropic principle that I had, after a lot of thought, concluded could not be the basis for a successful scientific theory. To see if we could do better, I formulated conditions that would allow a theory based on a landscape to be a real scientific theory. As an example, I had invented the CNS idea. This was all described in my book, *The Life of the Cosmos*.

Susskind's papers on these issues led me to revisit them, to see if anything that had happened since might change my mind. So I undertook a carefully argued paper on the AP and alternatives to it [a]. The dialogue with Lenny began when I sent a note to him, asking whether he might have any response to the arguments in that paper. At first there were some misunderstandings, because Susskind responded only to a summary and not the full paper. Nevertheless, some important points were raised, although nothing that requires modification of my original paper. This letter is my response to a paper Susskind put out in the course of our dialogue, making certain criticisms of cosmological natural selection (CNS) [b], and is mostly devoted to answering them.

We agree on several important things, among them that fundamental physics likely gives us a landscape of possible theories, while cosmology may give a multiverse containing a vast number of regions like our own universe. We disagree here mainly on one thing: the mechanism of reproduction we believe has been most important in populating the multiverse.

My main point is that *string theory will have much more explanatory power if the dominant mode of reproduction is through black holes, as is the case in the original version of CNS*. This is the key point I would hope to convince Susskind and his colleagues about, because I am sure that the case they want to make is very much weakened if they rely on the anthropic principle and eternal inflation.

Susskind believes instead that eternal inflation is the mode of reproduction. But suppose that everything Susskind wants to be true about both eternal inflation and the string-theory landscape turns out to be true. What is the best thing that could reasonably be expected to happen?

Weinberg, Vilenkin, Linde, and others proposed that in this case we might be able to explain the value of the vacuum energy, both during and after inflation. This is because it is the vacuum energy that determines how many universes are made in eternal inflation, and how large each one is.

However, a careful examination exposes two problems. The first is that the methods so far proposed to make predictions in this scenario are either logically flawed or ambiguous, so that the assumptions can be manipulated to get different predictions. This is explained in detail in section 5.1 of my paper. A second piece of bad news is that even if this can somehow be made to work, you can't expect to explain much more than the vacuum energy. The reason, as I explain in some detail in section 5.1.4, is that a statistical selection mechanism can only act to tune those parameters that strongly influence how many universes get created. As the selection mechanism in eternal inflation involves inflation, which happens at the grand unified scale, the low-energy parameters such as the masses of the light quarks and leptons are not going to have much of an effect on how many universes get created.

In order to tune the low-energy parameters, there must be a selection mechanism that is differentially sensitive to the parameters of low-energy physics. So we can ask, what possible mechanisms are there for production of universes within a multiverse, such that the number of universes made is sensitive to the values of light quark and lepton masses? I asked myself this question when I realized there would be a landscape of string theories.

The only answer I could come up with is reproduction through black holes. It works because a lot of low-energy physics and chemistry goes into the astrophysics that determines how many black holes get made.

Susskind complains that this is complicated, but it has to be complicated. The reason is that we are trying to understand a very curious fact, which is that, as noted by the people who invented the anthropic principle, the low-energy parameters seem tuned to produce carbon chemistry and long-lived stars. This is explained if CNS is true, because the formation of stars massive

Lee Smolin and Leonard Susskind

enough to become black holes depend on there being both carbon and a large hierarchy of stellar lifetimes.

Thus, if you like eternal inflation because it has a chance of explaining the tuning [of] the vacuum energy, you should like cosmological natural selection much more—because it has potentially much more explanatory power. It offers the only chance so far proposed to actually explain from string theory the parameters that govern low-energy physics. Also, as I argued in detail in my paper, the selection mechanism in CNS is falsifiable, whereas those proposed for eternal inflation so far are too ambiguous to lead to clean predictions.

Moreover, because the selection mechanism is dominated by known low-energy physics and chemistry, we really do know much more about it than about eternal inflation. We know the dynamics, we know the parameters, and we can use relatively well-tested astrophysical models to ask what the effect on the number of universes is of small changes in the parameters. None of this is true for inflation, where unfortunately there are a large variety of models which all are in agreement with observation but which give different predictions concerning eternal inflation.

Of course, it is possible that both mechanisms play a role. It might be useful to study this; so far no one has. It is premature to conclude, as Susskind does, that the production of universes by eternal inflation will dominate. Our universe has "only" 10^{18} black holes, but the total number of universes in CNS is vastly bigger than this, as there must have been a very large number of previous generations for the mechanism to work.

Susskind made a few direct criticisms of CNS which are easy to answer, as they have been considered earlier.

He raises the question of how many new universes are created per astrophysical black hole. In the initial formulation of

CNS, I presumed one, but some approximate calculations have suggested that the number could be variable. I discussed this in detail on page 320 of *Life of the Cosmos*. The reader can see the details there. What I concluded is that if theory predicts that the number of new universes created increases with the mass, by at least the first power of the mass, the theory can easily be disproved. This hasn't happened, but it could, and it is one of the ways CNS could be falsified. This is of course good not bad, for the more vulnerable a theory is to falsification, the better science it is, and the more likely we are to take it seriously if it nonetheless survives.

One of the assumptions of CNS is that the average change in the low-energy parameters when a new universe is created is small. Susskind says he doubts this is true in string theory. If Susskind is right, then CNS and string theory could not both be true. But I don't share his intuitions about this. I would have to invoke technicalities to explain why, but all that need be said here is that so far there are no calculations detailed enough to decide the issue. But there could be soon, as I mentioned before, using methods developed recently in loop quantum gravity. These methods may help us study what happens to singularities in string theory and may also provide a better framework to understand eternal inflation.

The rest of this note concerns Susskind's comments about black holes. He says, " . . . we have learned some things about black holes over the last decade that even Stephen Hawking agrees with. Black holes do not lose information." From this he draws the conclusion that "the quantum state of the offspring is completely unique and can have no memory of the initial state. That would preclude the kind of slow mutation rate envisioned by Smolin."

This is the central point, as Susskind is asserting that black holes cannot play the role postulated in CNS without contradicting the principles of quantum theory and results from string theory. I am sure he is wrong about this. I would like to carefully explain why. This question turns out to rest on key issues in the quantum theory of gravity, which many string theorists, coming from a particle physics background, have insufficiently appreciated.

The discussion about black holes "losing information" concerns processes in which a black hole forms and then evaporates. Hawking had conjectured in 1974 that information about the initial state of the universe is lost when this happens. Susskind and others have long argued that this cannot be true, otherwise the basic laws of quantum physics would break down.

As Hawking initially formulated the problem, the black hole would evaporate completely, leaving a universe identical to the initial one but with less information. This could indeed be a problem, but this is not the situation now under discussion. The present discussion is about cases in which a black-hole singularity has bounced, leading to the creation of a new region of space-time to the future of where the black-hole singularity would have been. In the future there are two big regions of space, the initial one and the new one. If this occurs, then some of the information that went into the black hole could end up in the new region of space. It would be "lost" from the point of view of an observer in the original universe, but not "destroyed," for it resides in the new universe or in correlations between measurements in the two universes.

The first point to make is that if this happens it does not contradict the laws of quantum mechanics. Nothing we know about quantum theory forbids a situation in which individual observers do not have access to complete information about the quantum state. Much of quantum information theory and quantum cryptography is about such situations. Generalizations of quantum theory that apply to such situations have been developed, and basic properties such as conservation of energy and probability are maintained. Using methods related to those developed in quantum information theory, Markopoulou and collaborators have shown how to formulate quantum cosmology so that it is sensible even if the causal structure is nontrivial, so that no observer can have access to all the information necessary to reconstruct the quantum state [c]. Information is never lost—but it is not always accessible to every observer.

So there is nothing to worry about: nothing important from quantum physics [d] is lost if baby universes are created in black holes and some information about the initial state of the universe ends up there.

A second point is that there is good reason to believe that in quantum gravity, information accessible to local observers decoheres in any case, because of the lack of an ideal clock. In particle physics, time is treated in an ideal manner and the clock is assumed to be outside of the quantum system studied. But when we apply quantum physics to the universe as a whole, we cannot assume this: The clock must be part of the system studied. As pointed out independently by Milburn [e] and by Gambini, Porto, and Pullin [f], this has consequences for the issue of loss of information. The reason is that quantum mechanical uncertainties come into the reading of the clock—so we cannot know exactly how much physical time is associated with the motion of the clock's hands. So if we ask what the quantum state is when the clock reads a certain time, there will be additional statistical uncertainties which grow with time. (In spite of this, energy

Lee Smolin and Leonard Susskind

and probability are both conserved.) But, as shown by Gambini, Porto, and Pullin, even using the best possible clock these uncertainties will dominate over any loss of information trapped in a black hole. This means that even if information is lost in black-hole evaporation, no one could do an experiment with a real physical clock that could show it.

I believe this answers the worries about quantum theory, but I haven't yet addressed Susskind's assertion that "we have learned some things about black holes over the last decade. Black holes do not lose information."

I've found that to think clearly and objectively about issues in string theory, it is necessary to first carefully distinguish conjectures from the actual results. Thus, over the last few years I've taken the time to carefully read the literature and keep track of what has actually been shown about the key conjectures of string theory. The results are described in two papers [g].

In this case, I'm afraid it is simply not true that the actual results in string theory—as opposed to so-far-unproven conjectures—support Susskind's assertions [h].

There are two classes of results relevant for quantum black holes in string theory. One concerns the entropy of very special black holes, which have close to the maximal possible charge or angular momenta for black holes. For this limited class of black holes the results are impressive, but it has not, almost ten years later, been possible to extend them to typical black holes. The black holes that were successfully described by string theory have a property that typical astrophysical black holes do not have: They have positive specific heat. This means that when you put in energy the temperature goes up. But most gravitationally bound systems, and most black holes, have the opposite property: You put in energy and they get colder. It appears that the methods used so far in string theory only apply to systems with positive specific heat, therefore no conclusions can be drawn for typical astrophysical black holes.

A second set of results concerns a conjecture by Maldacena. According to it, string theory in a spacetime with negative cosmological constant is conjectured to be equivalent to a certain ordinary quantum system, with no gravity. (That ordinary system is a certain version of what is called a gauge theory, which is a kind of generalization of electromagnetism.) Even if Maldacena's conjecture is true, that is no reason to assume there could not be baby universes where information was kept apart from an observer in the initial universe for a very long, but not infinite, time. This can be accomplished as long as all the different regions eventually come into causal contact so that, if one waits an infinite time, it becomes possible to receive the information that has gone into the baby universes.

But in any case Maldacena's conjecture has so far not been proven. There is quite a lot of evidence showing there is some relation between the two theories, but all of the results so far are consistent with a far weaker relationship holding between the two theories than the full equivalence Maldacena conjectured. This weaker relationship was originally formulated in a paper by Witten, shortly after the one of Maldacena. Except for a few special cases, which can be explained by special symmetry arguments, all the evidence is consistent with Witten's weaker conjecture. We should here recall a basic principle of logic that when a collection of evidence is explained by two hypotheses, one stronger and one weaker, only the weaker one can be taken to be supported by the evidence.

But Witten's conjecture requires only that there be a partial and approximate correspondence between the two theories. It does not forbid either baby universes or the loss of information by black holes. For example, Witten shows how some black holes can be studied using results in the other theory, but again it turns out that these are atypical black holes, with positive specific heat.

This discussion is related to a conjecture called the holographic principle (HP), an idea proposed by 't Hooft (and a bit earlier Crane) that Susskind brought into string theory. Susskind proposes a strong form of the HP, which holds that a complete description of a system resides in the degrees of freedom on its boundary. He takes Maldacena's conjecture as a demonstration of it. I believe here also the evidence better supports a weaker form (proposed with Markopoulou), according to which there is a relation between area and information but no necessity that the boundary has a complete description of its interior [i].

I would urge a similar caution with respect to Susskind's claim: "As repeatedly emphasized by 't Hooft, black holes are the natural extension of the elementary particle spectrum. This is especially clear in string theory where black holes are simply highly excited string states. Does that mean that we should count every particle as a black hole?"

As I mentioned, the only results in string theory that describe black holes in any detail describe only very atypical black holes. In those cases, they are related—at least by an indirect argument—to states described by string theory, but they are not in fact excitations of strings. They involve instead objects called D-branes. So Susskind must mean by "a highly excited string state" any state of string theory. But in this case the argument has no force, as stars, planets, and people must also be "highly excited string states." In any case, until there are detailed descriptions of typical black holes in string theory, it is premature to judge whether Susskind and 't Hooft have conjectured correctly.

Susskind attempts to invoke Hawking's authority here, and it is true that Hawking has announced that he has changed his view on this subject. But he has not yet put out a paper, and the transcript of the talk he gave recently doesn't provide enough details to judge how seriously we should take his change of opinion.

Next, Susskind refers to a paper by Horowitz and Maldacena, of which he says that "The implication is that if there is any kind of universe creation in the interior of the black hole, the quantum state of the offspring is completely unique and can have no memory of the initial state. That would preclude the kind of slow mutation rate envisioned by Smolin."

I read that paper and had some correspondence with its authors about it; unfortunately, Susskind misstates its implications. In fact that paper does not show that there is no loss of information, it merely assumes it and proposes a mechanism—which the authors acknowledge is speculative and not derived from theory—that might explain how it is that information is not lost. They do not show that information going into baby universes is precluded; in fact Maldacena wrote to me that "If black hole singularities really bounce into a second large region, I also think our proposal is false [j]."

Finally, Susskind suggests that loop quantum gravity will be inconsistent unless it agrees with his conjectures about black holes. I should then mention that there are by now sufficient rigorous results (reviewed in [k]) to establish the consistency of the description of quantum geometry given by loop quantum gravity. Whether it applies to nature is an open question, as is what it has to say about black hole singularities, but progress in both directions is steady.

Let me close with something Susskind and I agree about—which I learned from him back in graduate school: an idea called string/gauge duality according to which gauge fields, like those in electromagnetism and QCD have an equivalent description in terms of extended objects. For Susskind, those extended objects are strings. I believe that may be true at some level of approximation, but the problem is that we only know how to make sense of string theory in a context in which the geometry of spacetime is kept classical—giving a background in which the strings move.

But general relativity teaches us that spacetime cannot be fixed; it is as dynamical as any other field. So a quantum theory of gravity must be background-independent. We should then ask if there is a version of this duality in which there is no fixed, classical background, so that the geometry of spacetime can be treated completely quantum mechanically. Indeed there is; it is loop quantum gravity. Moreover, a recent uniqueness theorem [1] shows essentially that any consistent background-independent version of this duality will be equivalent to loop quantum gravity. For this reason, I believe it is likely that if string theory is not altogether wrong, sooner or later it will find a more fundamental formulation in the language of loop quantum gravity.

Indeed, what separates us on all these issues is the question of whether the quantum theory of gravity is to be background-independent or not. Most string theorists have yet to fully take on board the lesson from Einstein's general theory of relativity; their intuitions about physics are still expressed in terms of things moving in fixed background spacetimes. For example, the view of time evolution that Susskind wants to preserve is tied to the existence of a fixed background. This leads him to propose a version of the holographic principle which can only be formulated in terms of a fixed background. The strong form of Maldacena's conjecture posits that quantum gravity is equivalent to physics on a fixed background. The approaches string theory takes to black holes only succeed partially, because they describe black holes in terms of objects in a fixed background. Eternal inflation is also a background-dependent theory; indeed, some of its proponents have seen it as a return to an eternal, static universe.

On the other hand, those who have concentrated on quantum gravity have learned, from loop quantum gravity and other approaches, how to do quantum spacetime physics in a background-independent way. After the many successful calculations which have been done, we have gained a new and different intuition about physics, and it leads to different expectations for each of the issues we have been discussing. There is still more to do, but it is clear there need be—and can be—no going back to a pre–Einsteinian view of space and time. Anyone who still wants to approach the problems of physics by discussing how things move in classical background spacetimes—whether those things are strings, branes or whatever—are addressing the past rather than the future of our science.

References:
[a] Lee Smolin, Scientific alternatives to the anthropic principle, hep-th/0407213.
[b] Leonard Susskind, Cosmic natural selection, hep-th/0407266.
[c] E. Hawkins, F. Markopoulou, H. Sahlmann, Evolution in quantum causal histories, hep-th/0302111.
[d] In particular, global unitarity is automatically present whenever there is a global time coordinate, but need not be if that condition is not met. Quantum information accessible to local observables is propagated in terms of density matrices following rules that conserve energy and probability, because a weaker property, described in terms of completely positive maps, is maintained.
[e] G. J. Milburn, *Phys. Rev* A44, 5401 (1991).

[f] Rodolfo Gambini, Rafael Porto, Jorge Pullin, Realistic clocks, universal decoherence and the black hole information paradox hep-th/0406260, gr-qc/0402118 and references cited there.

[g] L. Smolin, How far are we from the quantum theory of gravity? hep-th/0303185; M. Arnsdorf and L. Smolin, The Maldacena conjecture and Rehren duality, hep-th/0106073.

[h] This is one of several key cases in which conjectures, widely believed by string theorists have not so far been proven by the actual results on the table. Another key unproven conjecture concerns the finiteness of the theory.

[i] F. Markopoulou and L. Smolin, Holography in a quantum space-time, hep-th/9910146; L. Smolin, The strong and the weak holographic principles, hep-th/0003056.

[j] Juan Maldacena, email to me, 1 November 2003, used with permission.

[k] L. Smolin, An invitation to loop quantum gravity, hep-th/0408048.

[l] By Lewandowski, Okolow, Sahlmann and Thiemann, see p. 20 of the previous endnote.

12
Science Is Not About Certainty

Carlo Rovelli

Theoretical physicist, Professeur de classe exceptionelle,
Université de la Méditerranée, Marseille; author, *The First
Scientist: Anaximander and His Legacy*

INTRODUCTION by Lee Smolin

Carlo Rovelli is a leading contributor to quantum gravity who
has also made influential proposals regarding the foundation of
quantum mechanics and the nature of time. Shortly after re-
ceiving his PhD, he did work that made him regarded as one
of the three founders of the approach to quantum gravity called
loop quantum gravity—the other two being Abhay Ashtekar
and me. Over the last twenty-five years, he has made numerous
contributions to the field, the most important of which developed the spacetime approach to quantum gravity called spin-
foam models. These have culminated over the last five years
in a series of discoveries that give strong evidence that loop
quantum gravity provides a consistent and plausible quantum
theory of gravity.

Rovelli's textbook, *Quantum Gravity,* has been the main introduction to the field since its publication in 2004, and his research group in Marseille has been a major center for incubating
and developing new talent in the field in Europe. Carlo Rovelli's
approach to the foundations of quantum mechanics is called relational quantum theory. He also, with the mathematician Alain
Connes, has proposed a mechanism by which time could emerge

from a timeless world—a mechanism called the thermal time hypothesis.

Science Is Not About Certainty

We teach our students: We say that we have some theories about science. Science is about hypothetico-deductive methods; we have observations, we have data, data require organizing into theories. So then we have theories. These theories are suggested or produced from the data somehow, then checked in terms of the data. Then time passes, we have more data, theories evolve, we throw away a theory, and we find another theory that's better, a better understanding of the data, and so on and so forth.

This is the standard idea of how science works, which implies that science is about empirical content; the true, interesting, relevant content of science is its empirical content. Since theories change, the empirical content is the solid part of what science is.

Now, there's something disturbing, for me, as a theoretical scientist, in all this. I feel that something is missing. Something of the story is missing. I've been asking myself, "What is this thing missing?" I'm not sure I have the answer, but I want to present some ideas on something else that science is.

This is particularly relevant today in science, and particularly in physics, because—if I'm allowed to be polemical—in my field, fundamental theoretical physics, for thirty years we have failed. There hasn't been a major success in theoretical physics in the last few decades after the standard model, somehow. Of course there are ideas. These ideas might turn out to be right. Loop quantum gravity might turn out to be right, or not. String theory might turn out to be right, or not. But we don't know, and for the moment Nature has not said yes, in any sense.

I suspect that this might be in part because of the wrong ideas we have about science, and because methodologically we're doing something wrong—at least in theoretical physics, and perhaps also in other sciences. Let me tell you a story to explain what I mean. The story is an old story about my latest, greatest passion outside theoretical physics—an ancient scientist, or so I say even if often he's called a philosopher: Anaximander. I'm fascinated by this character, Anaximander. I went into understanding what he did, and to me he's a scientist. He did something that's very typical of science and shows some aspect of what science is. What is the story with Anaximander? It's the following, in brief:

Until Anaximander, all the civilizations of the planet—everybody around the world—thought the structure of the world was the sky over our heads and the earth under our feet. There's an up and a down, heavy things fall from the up to the down, and that's reality. Reality is oriented up and down; Heaven's up and Earth is down. Then comes Anaximander and says, "No, it's something else. The Earth is a finite body that floats in space, without falling, and the sky is not just over our head, it's all around."

How did he get this? Well, obviously, he looked at the sky. You see things going around—the stars, the heavens, the moon, the planets, everything moves around and keeps turning around us. It's sort of reasonable to think that below us is nothing, so it seems simple to come to this conclusion. Except that nobody else came to this conclusion. In centuries and centuries of ancient civilizations, nobody got there. The Chinese didn't get there until the 17th century, when Matteo Ricci and the Jesuits went to China and told them. In spite of centuries of the Imperial Astronomical Institute, which was studying the sky. The Indians learned this only when the Greeks arrived to tell them. In Africa, in America, in Australia—nobody else arrived at this simple realization that the sky is not just over our head, it's also under our feet. Why?

Because obviously it's easy to suggest that the Earth floats in nothing, but then you have to answer the question, Why doesn't it fall? The genius of Anaximander was to answer this question. We know his answer—from Aristotle, from other people. He doesn't answer this question, in fact: He questions this question. He asks, "Why should it fall?" Things fall toward the Earth. Why should the Earth itself fall? In other words, he realizes that the obvious generalization—from every heavy object falling to the Earth itself falling—might be wrong. He proposes an alternative, which is that objects fall toward the Earth, which means that the direction of falling changes around the Earth.

This means that up and down become notions relative to the Earth. Which is rather simple to figure out for us now: We've learned this idea. But if you think of the difficulty when we were children of understanding how people in Sydney could live upside-down, clearly this required changing something structural in our basic language in terms of which we understand the world. In other words, "up" and "down" meant something different before and after Anaximander's revolution.

He understands something about reality essentially by changing something in the conceptual structure we use to grasp reality. In doing so, he isn't making a theory; he understands something that, in some precise sense, is forever. It's an uncovered truth, which to a large extent is a negative truth. He frees us from prejudice, a prejudice that was ingrained in our conceptual structure for thinking about space.

Why do I think this is interesting? Because I think this is what happens at every major step, at least in physics; in fact, I think this is what happened at every step in physics, not necessarily major. When I give a thesis to students, most of the time the problem I give for a thesis is not solved. It's not solved because the solution of the question, most of the time, is not in solving the question, it's in questioning the question itself. It's realizing that in the way the problem was formulated there was some implicit prejudice or assumption that should be dropped.

If this is so, then the idea that we have data and theories and then we have a rational agent who constructs theories from the data using his rationality, his mind, his intelligence, his conceptual structure doesn't make any sense, because what's being challenged at every step is not the theory, it's the conceptual structure used in constructing the theory and interpreting the data. In other words, it's not by changing theories that we go ahead but by changing the way we think about the world.

The prototype of this way of thinking—the example that makes it clearer—is Einstein's discovery of special relativity. On the one hand, there was Newtonian mechanics, which was extremely successful with its empirical content. On the other hand, there was Maxwell's theory, with its empirical content, which was extremely successful, too. But there was a contradiction between the two.

If Einstein had gone to school to learn what science is, if he had read Kuhn, and the philosophers explaining what science is, if he was any one of my colleagues today who are looking for a solution of the big problem of physics today, what would he do? He would say, "OK, the empirical content is the strong part of the theory. The idea in classical mechanics that velocity is relative: forget about it. The Maxwell equations: forget about them. Because this is a volatile part of our knowledge. The theories themselves have to be changed, OK? What we keep solid is the data, and we modify the theory so that it makes sense coherently, and coherently with the data."

That's not *at all* what Einstein does. Einstein does the contrary.

Carlo Rovelli

He takes the theories very seriously. He believes the theories. He says, "Look, classical mechanics is so successful that when it says that velocity is relative, we should take it seriously, and we should believe it. And the Maxwell equations are so successful that we should believe the Maxwell equations." He has so much trust in the theory itself, in the qualitative content of the theory—that qualitative content that Kuhn says changes all the time, that we learned not to take too seriously—and he has so much in that that he's ready to do what? To force coherence between the two theories by challenging something completely different, which is something that's in our head, which is how we think about time.

He's changing something in common sense—something about the elementary structure in terms of which we think of the world—on the basis of trust of the past results in physics. This is exactly the opposite of what's done today in physics. If you read *Physical Review* today, it's all about theories that challenge completely and deeply the content of previous theories, so that theories in which there's no Lorentz invariance, which are not relativistic, which are not general covariant, quantum mechanics, might be wrong. . . .

Every physicist today is immediately ready to say, "OK, all of our past knowledge about the world is wrong. Let's randomly pick some new idea." I suspect that this is not a small component of the long-term lack of success of theoretical physics. You understand something new about the world either from new data or from thinking deeply on what we've already learned about the world. But thinking means also accepting what we've learned, challenging what we think, and knowing that in some of the things we think, there may be something to modify.

What, then, are the aspects of doing science that I think are undervalued and should come up front? First, science is about constructing visions of the world, about rearranging our conceptual structure, about creating new concepts which were not there before, and even more, about changing, challenging, the *a priori* that we have. It has nothing to do with the assembling of data and the ways of organizing the assembly of data. It has everything to do with the way we think, and with our mental vision of the world. Science is a process in which we keep exploring ways of thinking and keep changing our image of the world, our vision of the world, to find new visions that work a little bit better.

In doing that, what we've learned in the past is our main ingredient—especially the negative things we've learned. If we've learned that the Earth is not flat, there will be no theory in the future in which the Earth is flat. If we have learned that the Earth is not at the center of the universe, that's forever. We're not going to go back on this. If you've learned that simultaneity is relative, with Einstein, we're not going back to absolute simultaneity, like many people think. Thus when an experiment measures neutrinos going faster than light, we should be suspicious and, of course, check to see whether there's something very deep that's happening. But it's absurd when everybody jumps and says, "OK, Einstein was wrong," just because a little anomaly indicates this. It never works like that in science.

The past knowledge is always with us, and it's our main ingredient for understanding. The theoretical ideas that are based on "Let's imagine that this may happen, because why not?" are not taking us anywhere.

I seem to be saying two things that contradict each other. On the one hand, we trust our past knowledge, and on the other hand, we are always ready to modify, in depth, part of our conceptual structure of the world. There's no contradiction between the two; the idea of the contradiction comes from what I see as the deepest misunderstanding about science, which is the idea that science is about certainty.

Science is not about certainty. Science is about finding the most reliable way of thinking at the present level of knowledge. Science is extremely reliable; it's not certain. In fact, not only is it not certain, but it's the lack of certainty that grounds it. Scientific ideas are credible not because they are sure but because they're the ones that have survived all the possible past critiques, and they're the most credible because they were put on the table for everybody's criticism.

The very expression "scientifically proven" is a contradiction in terms. There's nothing that is scientifically proven. The core of science is the deep awareness that we have wrong ideas, we have prejudices. We have ingrained prejudices. In our conceptual structure for grasping reality, there might be something not appropriate, something we may have to revise to understand better. So at any moment we have a vision of reality that is effective, it's good, it's the best we have found so far. It's the most credible we have found so far; it's mostly correct.

But, at the same time, it's not taken as certain, and any element of it is *a priori* open for revision. Why do we have this continuous . . . ? On the one hand, we have this brain, and it has evolved for millions of years. It has evolved for us, basically, for running across the savannah, for running after and eating deer and trying not to be eaten by lions. We have a brain tuned to meters and hours, which is not particularly well-tuned to think about atoms and galaxies. So we have to overcome that.

At the same time, I think we have been selected for going out of the forest, perhaps going out of Africa, for being as smart as possible, as animals that escape lions. This continuing effort on our part to change our way of thinking, to readapt, is our nature.

We're not changing our mind outside of nature; it's our natural history that continues to change us.

If I can make a final comment about this way of thinking about science, or two final comments: One is that science is not about the data. The empirical content of scientific theory is not what's relevant. The data serve to suggest the theory, to confirm the theory, to disconfirm the theory, to prove the theory wrong. But these are the tools we use. What interests us is the content of the theory. What interests us is what the theory says about the world. General relativity says spacetime is curved. The data of general relativity are that the Mercury perihelion moves 43 degrees per century with respect to that computed with Newtonian mechanics.

Who cares? Who cares about these details? If that were the content of general relativity, general relativity would be boring. General relativity is interesting not because of its data but because it tells us that as far as we know today, the best way of conceptualizing spacetime is as a curved object. It gives us a better way of grasping reality than Newtonian mechanics, because it tells us that there can be black holes, because it tells us there's a Big Bang. This is the content of the scientific theory. All living beings on Earth have common ancestors. This is a content of the scientific theory, not the specific data used to check the theory.

So the focus of scientific thinking, I believe, should be on the content of the theory—the past theory, the previous theories—to try to see what they hold concretely and what they suggest to us for changing in our conceptual frame.

The final consideration regards just one comment about this understanding of science, and the long conflict across the centuries between scientific thinking and religious thinking. It is often misunderstood. The question is, Why can't we live happily together and why can't people pray to their gods and study the universe without this continual clash? This continual clash is a little unavoidable, for the opposite reason from the one often presented. It's unavoidable not because science pretends to know the answers. It's the other way around, because scientific thinking is a constant reminder to us that we *don't* know the answers. In religious thinking, this is often unacceptable. What's unacceptable is not a scientist who says, "I know . . ." but a scientist who says, "I don't know, and how could you know?" Many religions, or some religions, or some ways of being religious, are based on the idea that there should be a truth that one can hold onto and not question. This way of thinking is naturally disturbed by a way of thinking based on continual revision, not just of theories but of the core ground of the way in which we think.

So, to sum up, science is not about data; it's not about the empirical content, about our vision of the world. It's about overcoming our own ideas and continually going beyond common sense. Science is a continual challenging of common sense, and the core of science is not certainty, it's continual uncertainty—I would even say, the joy of being aware that in everything we think, there are probably still an enormous amount of prejudices and mistakes, and trying to learn to look a little bit beyond, knowing that there's always a larger point of view to be expected in the future.

We're very far from the final theory of the world, in my field, in physics—extremely far. Every hope of saying, "Well we're almost there, we've solved all the problems" is nonsense. And we're wrong when we discard the value of theories like quantum mechanics, general relativity—or special relativity, for that matter— and try something else randomly. On the basis of what we know, we should learn something more, and at the same time we should somehow take our vision for what it is—a vision that's the best vision we have, but one we should continually evolve.

If science works, or in part works, in the way I've described, this is strongly tied to the kind of physics I do. The way I view the present situation in fundamental physics is that there are different problems: One is the problem of unification, of providing a theory of everything. The more specific problem, which is the problem in which I work, is quantum gravity. It's a remarkable problem because of general relativity. Gravity is spacetime; that's what we learned from Einstein. Doing quantum gravity means understanding what quantum spacetime is. And quantum spacetime requires some key change in the way we think about space and time.

Now, with respect to quantum gravity, there are two major research directions today, which are loops, the one in which I work, and strings. These are not just two different sets of equations; they are based on different philosophies of science, in a sense. The one in which I work is very much based on the philosophy I have just described, and that's what has forced me to think about the philosophy of science.

Why? Because the idea is the following: The best of what we know about spacetime is what we know from general relativity. The best of what we know about mechanics is what we know from quantum mechanics. There seems to be a difficulty in attaching the two pieces of the puzzle together: They don't fit well. But the difficulty might be in the way we face the problem. The best information we have about the world is still contained in these two theories, so let's take quantum mechanics as seriously as possible, believe it as much as possible. Maybe enlarge it a little bit to make it general relativistic, or whatever. And let's take general relativity as seriously as possible. General relativity has peculiar

Carlo Rovelli

features, specific symmetries, specific characteristics. Let's try to understand them deeply and see whether as they are, or maybe just a little bit enlarged, a little bit adapted, they can fit with quantum mechanics to give us a theory—even if the theory that comes out contradicts something in the way we think.

That's the way quantum gravity—the way of the loops, the way I work, and the way other people work—is being developed. This takes us in one specific direction of research, a set of equations, a way of putting up the theory. String theory has gone in the opposite direction. In a sense, it says, "Well, let's not take general relativity too seriously as an indication of how the universe works." Even quantum mechanics has been questioned, to some extent. "Let's imagine that quantum mechanics has to be replaced by something different. Let's try to guess something completely new"—some big theory out of which, somehow, the same empirical content of general relativity and quantum mechanics comes out in some limit.

I'm distrustful of this huge ambition, because we don't have the tools to guess this immense theory. String theory is a beautiful theory. It might work, but I suspect it's not going to work. I suspect it's not going to work because it's not sufficiently grounded in everything we know so far about the world, and especially in what I perceive as the main physical content of general relativity.

String theory's big guesswork. Physics has never been guesswork; it's been a way of unlearning how to think about something and learning about how to think a little bit differently by reading the novelty into the details of what we already know. Copernicus didn't have any new data, any major new idea; he just took Ptolemy, the details of Ptolemy, and he read in the details of Ptolemy the fact that the equants, the epicycles, the deferents, were in certain proportions. It was a way to look at the same construction from a slightly different perspective and discover that the Earth is not the center of the universe.

Einstein, as I said, took seriously both Maxwell's theory and classical mechanics in order to get special relativity. Loop quantum gravity is an attempt to do the same thing: take general relativity seriously, take quantum mechanics seriously, and out of that, bring them together, even if this means a theory where there's no time, no fundamental time, so that we have to rethink the world without basic time. The theory, on the one hand, is conservative because it's based on what we know. But it's totally radical, because it forces us to change something big in our way of thinking.

String theorists think differently. They say, "Well, let's go out to infinity, where somehow the full covariance of general relativity is not there. There we know what time is, we know what space is, because we're at asymptotic distances, at large distances." The theory is wilder, more different, newer, but in my opinion it's more based on the old conceptual structure. It's attached to the old conceptual structure and not attached to the novel content of the theories that have proven empirically successful. That's how my way of reading science coincides with the specifics of the research work that I do—specifically, loop quantum gravity.

Of course, we don't know. I want to be very clear. I think string theory is a great attempt to go ahead, by great people. My only polemical objection to string theory is when I hear—but I hear it less and less now—"Oh, we know the solution already; it's string theory." That's certainly wrong, and false. What's true is that it is a good set of ideas; loop quantum gravity is another good set of ideas. We have to wait and see which one of these theories turns out to work and, ultimately, be empirically confirmed.

Carlo Rovelli

This takes me to another point, which is, Should a scientist think about philosophy or not? It's the fashion today to discard philosophy, to say now that we have science, we don't need philosophy. I find this attitude naïve, for two reasons. One is historical. Just look back. Heisenberg would have never done quantum mechanics without being full of philosophy. Einstein would have never done relativity without having read all the philosophers and having a head full of philosophy. Galileo would never have done what he did without having a head full of Plato. Newton thought of himself as a philosopher and started by discussing this with Descartes and had strong philosophical ideas.

Even Maxwell, Boltzmann—all the major steps of science in the past were done by people who were very aware of methodological, fundamental, even metaphysical questions being posed. When Heisenberg does quantum mechanics, he is in a completely philosophical frame of mind. He says that in classical mechanics there's something philosophically wrong, there's not enough emphasis on empiricism. It is exactly this philosophical reading that allows him to construct that fantastically new physical theory, quantum mechanics.

The divorce between this strict dialogue between philosophers and scientists is very recent, in the second half of the 20th century. It has worked because in the first half of the 20th century people were so smart. Einstein and Heisenberg and Dirac and company put together relativity and quantum theory and did all the conceptual work. The physics of the second half of the century has been, in a sense, a physics of application of the great ideas of the people of the '30s—of the Einsteins and the Heisenbergs.

When you want to apply these ideas, when you do atomic physics, you need less conceptual thinking. But now we're back to basics, in a sense. When we do quantum gravity, it's not just application. The scientists who say "I don't care about philosophy" —it's not true that they don't care about philosophy, because they *have* a philosophy. They're using a philosophy of science. They're applying a methodology. They have a head full of ideas about what philosophy they're using; they're just not aware of them and they take them for granted, as if this were obvious and clear, when it's far from obvious and clear. They're taking a position without knowing that there are many other possibilities around that might work much better and might be more interesting for them.

There is narrow-mindedness, if I may say so, in many of my colleagues who don't want to learn what's being said in the philosophy of science. There is also a narrow-mindedness in a lot of areas of philosophy and the humanities, whose proponents don't want to learn about science—which is even more narrow-minded. Restricting our vision of reality today to just the core content of science or the core content of the humanities is being blind to the complexity of reality, which we can grasp from a number of points of view. The two points of view can teach each other and, I believe, enlarge each other.

13
The Energy of Empty Space That Isn't Zero

Lawrence Krauss

Physicist, cosmologist; director, Origins Project, Arizona State University; author, *A Universe from Nothing*

I just returned from the Virgin Islands, from a delightful event—a conference in St. Thomas—that I organized with twenty-one physicists. I like small events, and I got to hand-pick the people. The topic of the meeting was "Confronting Gravity." I wanted to have a meeting where people would look forward to the key issues facing fundamental physics and cosmology. And if you think about it, those issues all revolve in one way or another around gravity.

Someone at the meeting said, "Well, you know, don't we understand gravity? Things fall." But really, many of the key ideas that right now are at the forefront of particle-physics cosmology relate to our lack of understanding of how to accommodate gravity and quantum mechanics.

I invited a group of cosmologists, experimentalists, theorists, and particle physicists and cosmologists. Stephen Hawking came. We had three Nobel laureates: Gerard 't Hooft, David Gross, and Frank Wilczek; well-known cosmologists and physicists, such as Jim Peebles at Princeton, Alan Guth at MIT, Kip Thorne at Caltech, Lisa Randall at Harvard; experimentalists, such as Barry Barish of LIGO, the gravitational-wave observatory. We had observational cosmologists, people looking at the cosmic microwave background; we had Maria Spiropulu, from CERN, who's working on the Large Hadron Collider—which a decade ago people wouldn't have thought was a probe of gravity, but now, due to recent work in the possibility of extra dimensions, it might be.

I wanted to have a series of sessions where we would, each of us, try and speak somewhat provocatively about what each of us was thinking about, what the key issues are, and then have a lot of open time for discussion. So the meeting was devoted with a lot of open time for discussion, a lot of individual time for discussion, as well as some fun things, like going down in a submarine, which we did. It was a delightful event, where we defied gravity by having buoyancy, I guess.

I came away from this meeting realizing that the search for gravitational waves may be the next frontier. For a long time, I pooh-poohed it in my mind, because clearly it's going to be a long time before we could ever detect them, if they're there. And it wasn't clear to me what we'd learn, except that they exist. But one of the key worries I have as a cosmologist right now is that we have these ideas, and these parameters, and every experiment is consistent with this picture, and yet nothing points to the fundamental physics beneath it.

It's been very frustrating for particle physicists—and some people might say it's led to sensory deprivation which has resulted in the hallucination otherwise known as string theory. And that could be true. But in cosmology what we're having now is this cockamamie universe. We've discovered a tremendous amount. We've discovered that the universe is flat, which most of us theorists thought we knew in advance, because it's the only beautiful universe. But why is it flat? It's full of not just dark matter but this crazy stuff called dark energy that no one understands. This was an amazing discovery, in 1998 or so.

What's happened since then is that every single experiment agrees with this picture without adding insight into where it comes from. Similarly, all the data are consistent with ideas from inflation and everything is consistent with the simplest predictions of that, but not in a way you can necessarily falsify. Everything is consistent with this dark energy that looks like a cosmological constant—which tells us nothing.

It's a little subtle, but I'll try and explain it.

We've got this weird antigravity in the universe which is making the expansion of the universe accelerate. Now, if you plug in the equations of general relativity, the only thing that can antigravitate is the energy of nothing. Now, this has been a problem in physics since I've been a graduate student. It was such a severe problem that we never talked about it. When you apply quantum mechanics and special relativity, empty space inevitably has energy. The problem is, [it has] way too much energy. It has 120 orders of magnitude more energy than is contained in everything we see!

Now, that's the worst prediction in all of physics. You might ask, "If that's such a bad prediction, then how do we know empty space can have energy?" The answer is, "We know empty space isn't empty, because it's full of these virtual particles that pop in and out of existence, and we know *that* because if you try and calculate the energy level in a hydrogen atom and you don't include those virtual particles, you get a wrong answer." One of the greatest developments in physics in the 20th century was to realize that when you incorporate special relativity into quantum mechanics, you have virtual particles that can pop in and out of existence and they change the nature of a hydrogen atom—because a hydrogen atom isn't just a proton and electron. That's the wrong picture, because every now and then you have an electron-positron pair that pops into existence. And the electron is going to want to hang around near the proton. Because it's oppositely charged, the positron is going to be pushed out to the outskirts of the atom, and while they're there they change the charged distribution in the atom in a very small but calculable way. Feynman and others calculated that effect, which allows us to get agreement between theory and observation at the level of nine decimal places. It's the best prediction in all of science. There's no other place in science where from fundamental principles you can calculate a number and compare it to an experiment at nine decimal places.

But then when we ask, "If they're there, how much should they contribute to the energy in the universe?" we come up with the *worst* prediction in physics. It says that if empty space has so much energy we shouldn't be here. And physicists like me, theoretical physicists, knew they had the answer. They didn't know how to get there. It reminds me of the Sidney Harris cartoon where you've got this big equation and the answer, and the middle step says, "And then a miracle occurs." And then one scientist says to the other, "I think you have to be a little more specific at this step right here."

The answer had to be zero: The energy of empty space had to be precisely zero.

Why? Because you've got these virtual particles that are apparently contributing huge amounts of energy, you can imagine in physics how underlying symmetries in nature can produce exact cancellations—that happens all the time. Symmetries produce two numbers that are exactly equal and opposite, because somewhere there's an underlying mathematical symmetry of equations. So you can understand how symmetries could somehow cause an exact cancellation of the energy of empty space. But what you couldn't understand was how to cancel a number to

120 decimal places and leave something finite left over. You can't take two numbers that are very large and expect them to almost exactly cancel, leaving something that's 120 orders of magnitude smaller left over. And that's what would be required to have an energy that was comparable with the observational upper limits on the energy of empty space.

We knew the answer: There was a symmetry, and the number had to be exactly zero. Well, what have we discovered? There appears to be this energy of empty space that *isn't* zero! This flies in the face of all conventional wisdom in theoretical particle physics. It's the most profound shift in thinking, perhaps the most profound puzzle, in the latter half of the 20th century. And it may be so for the first half of the 21st century, or maybe all the way to the 22nd century. Because, unfortunately, I happen to think we won't be able to rely on experiment to resolve this problem. When we look out at the universe, if this dark energy is something that isn't quite an energy of empty space but just something that's pretending to be that, we might measure that it's changing over time. Then we would know that the actual energy of empty space is really zero and this is some cockamamie thing that's pretending to be energy of empty space. Many people have hoped they'd see that, because then you wouldn't need quantum gravity, which is a theory we don't yet have, to understand this apparent dark energy.

Indeed, one of the biggest failures of string theory's many failures, I think, is that it never successfully addressed this cosmological-constant problem. You'd think, if you had a theory of quantum gravity, it would explain precisely what the energy of empty space should be. And we don't have any other theory that addresses that problem, either. But if this thing really isn't vacuum energy—if it's something else—then you might be able to find out what it is and learn and do physics without having to understand quantum gravity.

The problem is, when we actually look out, every measure we've made right now is completely consistent with a constant energy in the universe over cosmological time. And that's consistent with the cosmological constant—with vacuum energy. So if you make the measurement that it's consistent with that, you learn nothing. Because it doesn't tell you that it's vacuum energy, because there could be other things that could mimic it. The only observation that would tell you—give you positive information—is if you could measure that it was changing over time. Then you'd know it wasn't vacuum energy.

And if we keep measuring this quantity better and better and better, it's quite possible that we'll find out that it looks more and more like a vacuum energy, and we're going to learn nothing. The only way to resolve this problem will be to have a theory, and theories are a lot harder to come by than experiments. Good ideas are few and far between. What we're really going to need is a good idea, and it may require an understanding of quantum gravity or it may require that you throw up your hands, which is what we're learning that a lot of people are willing to do.

In the Virgin Islands we had a session on the anthropic principle, and what's surprising is how many physicists have said, "You know, maybe the answer is an anthropic one." Twenty years ago, if you'd asked physicists if they hoped that one day we'd have a theory that tells us why the universe is the way it is, you'd have heard a resounding "Yes." They would all say, "That's why I got into physics." They might paraphrase Einstein, who said—while referring to God, but not really meaning God—that the question that really interested him was: Did God have any choice in the creation of the universe? What he really meant by that was, Is there only one consistent set of laws that works? If you changed one, if you twiddled one aspect of physical reality, would it all fall apart? Or are there lots of possible viable physical realities?

Twenty years ago, most physicists would have said, on the basis of four hundred and fifty years of science, that they believed that there's only one allowed law of nature that works—that ultimately we might discover fundamental symmetries and mathematical principles that cause nature to be the way it is, because it's always worked that way.

So, that's the way science has worked. But now—because of this energy of empty space that's so inexplicable that if it really is an energy of empty space, the value of that number is so ridiculous that it's driven people to think that maybe, maybe, it's an accident of our environment, that physics is an environmental science, that certain fundamental constants in nature may just be accidents and there may be many different universes in which the laws of physics are different, and the reasons those constants have the values they have in our universe might be because we're there to observe them. . . .

This is not intelligent design. It's the opposite of intelligent design. It's a kind of cosmic natural selection. The qualities we have exist because we can survive in this environment. That's natural selection, right? If we couldn't survive, we wouldn't be around. Well, it's the same with the universe. We live in a universe—in this universe. We've evolved in this universe because this universe is conducive to life. There may be other universes that aren't conducive to life, and lo and behold, there isn't life in them. That's the kind of cosmic natural selection.

We're allowed to presume anything. The key question is, Is it a scientific question to presume there are other universes? That's something we were looking at in the meeting as well. I wrote a piece where I argued that it's a disservice to evolutionary theory to call string theory a theory, for example. Because it's clearly not a theory in the same sense that evolutionary theory is, or that quantum electrodynamics is, because those are robust theories that make rigorous predictions that can be falsified. And string theory is just a formalism now, which one day might be a theory. And when I'm lecturing, talking about science, and people say to me, "Evolution is just a theory," I say, "In science, 'theory' means a different thing," and they say, "What do you mean? Look at string theory, how can you falsify that? It's no worse than intelligent design."

I do think there are huge differences between string theory and intelligent design. People who are doing string theory are earnest scientists who are trying to come up with ideas that are viable. People who are doing intelligent design aren't doing any of that. But the question is, "Is it falsifiable?" And do we do a disservice to real theories by calling hypotheses, or formalisms, "theories"? Is a multiverse—in one form or another—science?

In my sarcastic moments, I've argued that the reason some string theorists have latched onto the landscape idea so much is that since string theory doesn't make any predictions, it's good to have a universe where you can't make any predictions. But less sarcastically, if you try and do science with that idea. . . . You can try and do real science, and calculate probabilities, but whatever you do, you find that all you get is suggestive arguments. Because if you don't have an underlying theory, you never know.

I say, "Well, what's the probability of our universe having a vacuum energy, if it's allowed to vary over different universes?" Then I come up with some result that's interesting—and Steven Weinberg was one of the first people to point out that if the value of the energy of empty space was much greater than it is, then galaxies wouldn't have formed, and astronomers wouldn't have formed. So that gave rise to the anthropic argument that, well, maybe that's why it [the vacuum energy] is what it is—it can't be much more.

But the problem is, you don't know if that's the only quantity that's varying. Maybe there are other quantities that are varying. Whatever you're doing is always a kind of ad-hoc suggestive thing, at best. You can debate it, but it doesn't lead very far. It's not clear to me that the landscape idea will be anything but impotent. Ultimately it might lead to interesting suggestions about things, but real progress will occur when we actually have new ideas. If string theory is the right direction—and I'm willing to argue that it might be, even if there's no evidence that it is right now—then a new idea that tells us a fundamental principle for how to turn that formalism to a theory will give us a direction that will turn into something fruitful. Right now, we're floundering. We're floundering in a lot of different areas.

As a theorist, when I go to meetings I often get much more out of the experimental talks, because I often know what's going on in theory—or at least I like to think I do. I was profoundly affected by the experimental talks. In principle, we're now able to be sensitive to gravitational waves that might change a meter stick that's 3 kilometers long by a length equal to less than the size of an atom! It's just amazing that we have the technology to do this. While that isn't actually detecting any gravitational waves, there's no technological obstructions to going to the advanced stage. Gravitational waves may, indeed, allow us a probe that might take us beyond our current state of having observations that don't lead anywhere. I was very impressed with these findings.

At the same time, we had a talk from Eric Adelberger at the University of Washington, who's been trying to measure Newton's law on small scales. You might think, "Who would want to measure Newton's law on small scales?" But one of the suggestions for extra dimensions is that on small scales, gravity has a different behavior. There has been some tantalizing evidence, which went through the rumor mills, that suggested that in these experiments in Seattle they were seeing evidence for deviations from Newton's theory. And Adelberger talked about some beautiful experiments. As a theorist, I'm just always amazed that they can even do these experiments. And he gave some new results. There are some tentative new results—which of course are not a surprise to me—that suggest that there's as yet no evidence for a deviation from Newton's theory.

Many of the papers in particle physics over the last five to seven years have been involved with the idea of extra dimensions of one sort or another. And while it's a fascinating idea, I have to say it's looking to me like it's not yet leading anywhere. The experimental evidence against it is combining with what I see as a theoretical diffusion—a breaking off into lots of parts. That's happened with string theory. I can see it happening with extra-dimensional arguments. We're seeing that the developments from this idea, which has captured the imaginations of many physicists, hasn't been compelling.

Right now, it's clear that what we really need is some good new ideas. Fundamental physics is at kind of a crossroads. The observations have just told us that the universe is crazy but haven't told us what direction the universe is crazy in. The theories have been incredibly complex and elaborate but haven't yet made any compelling inroads. That can either be viewed as depressing or exciting. For young physicists it's exciting in the sense that it means that the field is ripe for something new.

The great hope for particle physics, which may be a great

hope for quantum gravity, is the next large particle accelerator. We've gone thirty years without a fundamentally new accelerator that can probe a totally new regime of the subatomic world. We would have had it, if our legislators had not been so myopic. It's amazing to think that if they hadn't killed the Superconducting Super Collider it would have already been running for ten years. The Large Hadron Collider is going to come online next year, and one of two things could happen: It could either reveal a fascinating new window on the universe and a whole new set of phenomena that will validate or refute the current prevailing ideas in theoretical particle physics—supersymmetry etc.—or it might see absolutely nothing. I'm not sure which I'm rooting for. But it is at least a hope, finally, that we may get an empirical handle that will at least constrain the wild speculation that theorists like me might make.

Such a handle comes out of the impact of the recent cosmic-microwave-background studies on inflation theory. I read in the *New York Times* that Alan Guth was smiling, and Alan Guth was sitting next to me at the conference when I handed him the article. He was smiling, but he always smiles, so I didn't know what to make of it, but I think the results that came out of the cosmic microwave background studies were twofold. Indeed, as the *Times* suggested, they validate the notions of inflation. But I think that's just journalists searching for a story.

Because if you look at what quantitatively has come out of the new results, they're exactly consistent with the old results. Which also validate inflation. They reduce the error bars a little bit, by a factor of 2. I don't know if that's astounding. But what's intriguing to me is that while everything is consistent with the simplest [inflation] models, there's one area where there's a puzzle. On the largest scales, when we look out at the universe, there doesn't seem to be enough structure—not as much as inflation would predict. Now the question is, Is that a statistical fluke?

That is, we live in one universe, so we're a sample of one. With a sample of one, you have what's called a large sample variance. And maybe this just means we're lucky, that we just happen to live in a universe where the number is smaller than you'd predict. But when you look at the CMB map, you also see that the structure that's observed is in fact, in a weird way, correlated with the plane of the Earth around the sun. Is this Copernicus coming back to haunt us? That's crazy. We're looking out at the whole universe. There's no way there should be a correlation of structure with Earth's motion around the sun—the plane of the Earth around the sun, the ecliptic. That would say that we're truly the center of the universe.

The new results are either telling us that all of science is wrong and we're the center of the universe or maybe that the data are simply incorrect. Or maybe it's telling us there's something weird about the microwave-background results and maybe, maybe, there's something wrong with our theories on the larger scales. And, of course, as a theorist I'm certainly hoping it's the latter, because I want theory to be wrong, not right. Because if it's wrong, there's still work left for the rest of us.

Lawrence Krauss

14
Einstein: An *Edge* Symposium

Brian Greene, Walter Isaacson, Paul Steinhardt

Brian Greene: String theorist, Columbia University; author, *The Fabric of the Cosmos*

Walter Isaacson: President and CEO, The Aspen Institute; author, *Einstein: His Life and Universe*

Paul Steinhardt: Theoretical physicist and Albert Einstein Professor of Science, Princeton University; coauthor (with Neil Turok), *Endless Universe: Beyond the Big Bang*

INTRODUCTION by John Brockman

The coincidence in spring 2007 of Walter Isaacson's Einstein biography (*Einstein: His Life and Universe*) hitting the #1 spot on the *New York Times* bestseller list, coupled with the publication of *Endless Universe: Beyond The Big Bang,* by Paul Steinhardt and Neil Turok, created an interesting opportunity.

I invited Walter, Paul, and Columbia University string theorist Brian Greene to participate in an *Edge* symposium on Einstein. Walter, Paul, and Brian showed up for the session during the summer of 2007.

A year earlier, in *My Einstein,* a book of essays by twenty-four leading thinkers, I had asked each of the contributors to share their thoughts on who Einstein was to them. This led me to ask the same questions to the *Edge* symposium participants.

"I'd say my Einstein surrounded my learning—not learning, really; hearing—in junior high school that there's this feature of time whereby if you're moving relative to somebody else, time elapses at a different rate compared to the person who's stationary," Brian said. "And thinking to myself, 'That sounds completely nuts. I really want to understand what this is all about.' And little by little finally learning what it actually means, and going on from there to try to push the story a little bit further."

Walter's response: "Einstein is obviously my father, who as an engineer loved science and instilled that in me, but also has a lot of Einstein's moral nature to him, and political morality to him. I remember every day growing up, his asking me questions and pushing me in a certain way. One of the things I've learned as a biographer, and the first thing you learn, is that as you write about your subject, it's all about Dad: For Ben Franklin it's all living up to his father in a certain way. Even for Einstein, a bit—his father's an engineer. And then the second thing you learn is, even for the biographer it's all about Dad, and that's why I wanted to write about Einstein. I shouldn't say my father's an Einstein, he's just an engineer in New Orleans, but that was his aspirational secular saint, and so I wrote the book and dedicated it to my father."

Paul had a similar response: "One of my earliest memories of childhood was sitting on my father's knee and his telling me stories about scientists and discovery. He wasn't a scientist, he was a lawyer, but for some reason he used to tell me stories about scientists and different discoveries they made—I remember stories about Madame Curie and Einstein and others. From that very initial instance, what I wanted to do was be in a field where you got to make discoveries. The thing that always impressed me the more I learned about Einstein was his uncanny ability to take the wealth of phenomena that people were studying at the time, and pick out not only which were the important questions but which were the important questions that were answerable. There are always lots of questions you'd like to answer, but knowing whether or not you have the technology, the mathematical technology, and the right ideas to attack them at the time—that's a real talent. Einstein had the incredible talent to do that over and over and over again, ahead of any of his contemporaries. So, for me he's the ultimate discoverer. That is my Einstein."

Einstein: An *Edge* Symposium

BRIAN GREENE: When it comes to Albert Einstein, his contributions are of such incredible magnitude that to get inside his head, and even for a moment to get a feel for what it would be like to see the world with such clarity and such insight, would be amazing.

But if I was going to ask him one question, I would probably stick to one a little bit more down to earth, which is: He famously said that when it came to the general theory of relativity, in some sense he wasn't waiting for the data to show whether it was right or wrong; the theory was so beautiful that it just had to be right. And when the data came in and confirmed it, he claimed he wasn't even surprised, he in fact famously said that had the data turned out differently, he would have been sorry for the Dear Lord, because the theory was correct. That's how much faith he had in the theory.

So the question I have is: We many of us, are working on Einstein's legacy in a sense, which is trying to find the unified theory that he looked for such a long time and never found, and we've been pursuing an approach called superstring theory for many years now. And it's a completely theoretical undertaking. It's completely mathematical. It has yet to make contact with experimental data. I'd like to ask Einstein what he would think of this approach to unification. Does he see the same kind of beauty, the same kind of elegance, the same kind of powerful incisive ideas in this framework to give him the confidence that he had in the general theory of relativity?

It would be great to have a response from him in that regard, because we don't know when we're going to make contact with experimental data. I think most of us in the field absolutely will never have faith that this approach is right until we do make contact with data, but it would be great to have the insight of the Master as to whether he feels that this smells right—that it's going in the right direction. Many of us think it is, but it would be great to have his insight on that question as well.

WALTER ISAACSON: I was going to ask him the same question Brian asked him, but I'll extend it now a bit more. Einstein, in the final two decades of his life—and even the final two hours of his life, on his deathbed—is writing equations, very mathematical, trying to do the unified theory that will bring together the various forces of nature into a field-theory approach.

Brian posed the question of whether or not Einstein would approve of this—and I really think he would, because if you look at the maybe twelve serious efforts he made toward a unified theory, they do have so much in common with the mathematics and the mathematical approach that's being done by superstring theory, including looking at extra dimensions and using the mathematics that way, to try to find the elegant mathematical solution.

That would lead to the next question I have about Einstein, which is, in the first part of his career and, may I posit, the more successful part of his career, he didn't rely that much on mathematical formalism. Instead, in all of the 1905 papers and in the main thought experiments that set him on the way toward general relativity, culminating in 1915, he had some physical insight.

Brian Greene, Walter Isaacson, Paul Steinhardt

In fact, the people looking at his general theory of relativity call it the mathematical strategy and the physics strategy.

Obviously they're not totally separate, in their iterative process, but he spent the period from 1905 through at least 1914 almost disparaging mathematics as a clean-up act that people would come along and help him do, once he understood the equivalence of gravity and acceleration or the other great thought experiments he did.

If you look at what he does later in life, with the unified theory, people like Banesh Hoffmann and others who were his collaborators say we had no physical insight to guide us, nothing like the principle of the equivalence of gravity and acceleration, or some other great insight, and instead it became more and more mathematical formalism, without what Einstein called the ground lights that would touch us, as we've just said, to physical reality more. And there are some who think—and I kind of feel this way, which is why I've adopted this idea—he had used the physical strategy, the physics approach, so much from 1905 to around 1913-1914, and even in the Zurich notebooks where he tries to get general relativity and the equations of gravity right, and he just can't quite get them, and he's racing against David Hilbert, who's a Göttingen mathematician who has the advantage of being a better mathematician but also an added advantage of not being as good of a physicist. Hilbert's not there worrying about whether it reduces to the Newtonian in the weak field or whatever—he's just pursuing general covariance as a mathematical strategy in order to get the field equations of gravity. Einstein finally adopts that approach and it puts him there, it makes him succeed through what is a very mathematical strategy, and then for the rest of his life he spends a lot of time on mathematical formalism instead of worrying about the intuitive physics behind everything.

Was that the right approach? Is that what's happening with string theory? Is that the better way to do it; is that what you have to do? As Einstein said, when he was asked about this: That's the way you have to approach things now; we don't have any blinding new physics insight.

Finally the bigger question is, when he fights—and I do think his quest for a unified theory comes out of his discomfort with quantum mechanics—when he is pushing against the people in the realm of quantum mechanics, they push back, and they say things like, "Well, we're just doing what you used to do; we're questioning every assumption. We're saying that unless you can observe something there's no reason to posit that it exists." Einstein is saying, "Yes, but that doesn't make sense now." They respond, "Well, you always questioned authority and questioned everything unless you could actually observe it, and now you're resisting us."

Einstein said, "Well, to punish me for my contempt for authority, fate has turned me into an authority myself." I'm no longer quite as rebellious, is what he's saying. So why is it that he becomes in some ways more defensive of the classical order and less rebellious, even as he's trying to pursue the unified theory?

PAUL STEINHARDT: The question I have about Einstein relates to the one Brian raised, but with a twist, because I see what has been happening in theoretical physics in the last thirty years, and especially in the last few years, maybe from a slightly different point of view.

Over the last thirty years, there have been grand ideas that have emerged in theoretical physics that were meant to simplify our understanding of the universe. One is the idea of inflation— the idea that there was a period of very rapid expansion that smoothed and flattened the universe and which explains why, when we look out anywhere in space, it looks almost the same everywhere. And it is, of course, based on Einstein's general theory of relativity and relates directly to his introduction of the cosmological constant back in 1917, but elaborated in a way that this rapid expansion would only occur in the early universe and not in the later universe. Just to explain why the universe is the same, or looks the same, almost everywhere.

The other grand idea that has been developing is the one Brian has written so elegantly about, which is the idea of string theory—that everything in nature is made of quantum vibrating strings, and that we can derive a simple unified theory to explain all the physical laws that we see.

The hope was that string theory would explain the microphysical world and inflation would explain the macrophysical world.

What have we learned from these two grand theories, inflation and string theory? Well, in the last few years, especially in the last decade, we've learned that—at least to my way of reckoning things—neither of them is really delivering on their promise. It turns out inflation doesn't do what we originally thought it did back when it was introduced in the 1980s. It doesn't take an original initially complicated inhomogeneous, nonuniform, curved and warped universe and flatten it out everywhere—and leave it with a universe which is full of matter that we see, which is then smoothed and flat. Instead, what inflation does, once it takes hold, is continue to make a more and more inflating universe, only occasionally leaving behind a few pockets of universe that have matter we would recognize and that might be inhabitable.

And in fact among those pockets, it seems that the pockets that would look like the ones we observe would be exceedingly rare. So whereas inflation was designed to explain why the universe is as uniform as it is, and why that's a likely occurrence, it seems the theory is leading us to a point of view where with inflation we're unlikely to find pockets of the universe that look as smooth and flat as we observe.

Similarly, the hope for string theory was that it would uniquely explain why the laws of physics are what they are. But developments over the last few years suggest that string theory doesn't make a unique prediction for the physical laws—there might be a googol, or many googols, of possibilities. And the ones we happen to observe are not particularly likely—at least there's no reason why they should be likely.

So a key conclusion, according to the current view of string theory and inflationary theory, is that the fundamental nature of the universe is random. Although the universe seems to be remarkably uniform everywhere as far as we can see, our leading theories currently suggest that the conditions we observe are actually very rare and unlikely phenomena in the universe entire. And I wonder what Einstein would have thought about that. I wonder if he would have found that idea—that is, a theory of this type—to be acceptable. My own point of view is that we have to change one or both of these two key components of our understanding of the universe. We either have to dramatically revise them or we have to overhaul them entirely, replace them with something that combines to make a powerful theory that really does explain, in a powerful way, why the universe is the way that it is.

ISAACSON: Why is there such a personal theological argument—I can see it between the Sam Harrises and the Dawkinses and Christopher Hitchens versus those who are strong believers—that when people start debating string theory, their faces turn red.

Brian Greene, Walter Isaacson, Paul Steinhardt

STEINHARDT: I think it's for the reason that I was beginning to raise. In my view, and in the eyes of many others, fundamental theory has crashed at the moment. Instead of delivering what it was supposed to deliver—a simple explanation of why the masses of particles and their interactions are what they are—we get instead the idea that string theory allows googols of possibilities and there's no particular reason for the properties we actually observe. They have been selected by chance. In fact, most of the universe has different properties. So the question is, Is that a satisfactory explanation of the laws of physics? In my own view, if I had walked in the door with a theory not called string theory and said that it's consistent with the observed laws of nature but, by the way, it also gives a googol other possibilities, I doubt I would have been able to say another sentence. I wouldn't have been taken seriously.

GREENE: You really think that? I don't mean to get technical, but take the standard model of particle physics, which is the quantum field theory that people have developed over the course of a number of decades and that we generally view as the most solid, pinnacle achievement of particle physics. When you look at the framework within which the standard model of particle physics sits—namely, relativistic quantum field theory—you do find that there are a googol, if not more, possible universes that that framework is capable of describing. The masses of the particles can be changed arbitrarily and the theory still makes sense, it's internally consistent; you can change the strengths of the forces, the strengths of the coupling constants.

So if you see the standard model, the one we all think is so spectacular, within the landscape of theories that that framework can give rise to, it seems to me that when you walk in the door and you say, "I've got this theory called the standard model to describe all physics," isn't everybody excited? If I were to use the same benchmark for judging it, you'd think I would ignore it as well, since it actually is part of a family of theories that can describe a googol of different universes. So how is that any worse than string theory?

STEINHARDT: Well, I think there's a key difference, which is that no one believes that the standard model is the ultimate theory, and string theory is claiming to be the ultimate theory.

GREENE: Oh, I think we should put claims of that sort—

STEINHARDT: But the question was raised, Why are people upset about it? And the answer is, Because whether you personally believe it or not, string theory has been advertised as being the ultimate theory from which we should be able to understand—

GREENE: I guess I would say it's unfortunate that people get worked up over that kind of advertising. If you look at the history of string theory, I agree with you; there was a time when people thought this could be it—the final theory that would describe everything. In fact, it still may be it.

But I think there was a certain kind of youthful exuberance that took hold when the theory was in its early infancy in the 1980s and early '90s and so forth, which perhaps was a little bit unfounded, because it was such an immature theory that you really couldn't make pronouncements about it that you could have any real faith in. I personally—as do, I think, many string theorists—view string theory as a possible next step toward a deeper understanding of the laws of physics. It could be the final step; we can't judge yet.

But I think the most sober way of looking at it is that we have quantum mechanics, we have general relativity, we have to put them together in some consistent way. String theory is a possible way of doing that, and therefore we should explore it and see where it goes. I think it would be unfortunate if simply by virtue of its being advertised within a certain framework of it being the final theory, one then judges it differently from any other scientific theory, which is on its merits.

STEINHARDT: That's a stupendous retreat from what many people have claimed.

GREENE: You really think so?

STEINHARDT: Yes, sure. And it's worse than that. Some people even claim this idea that you have this googol, or perhaps infinite, number of possibilities is something we should come to accept—that it's now derived from string theory, that string theory should be accepted as true, and since it has led to this multiplicity of possibilites, we should all accept this conclusion as true.

GREENE: Naturally, scientists quite generally, and string theorists in particular, often describe their work without giving all of the associated qualifications all of the time. I, for example, have spoken of string theory as a possible final theory, as the possible theory that would unite all forces and all matter in one consistent framework. And I generally try to say—but perhaps not always—that this is not yet a proven theory; this is our hope for what it will achieve. We aren't certain that this is where it's going to lead. We just need to explore and see where we land.

Similarly, I think that if you sat down and spoke to the folks you're referring to in a more informal setting—who talk about having all of these different universes emerge from string theory and about how it's a new framework that we have to think about things, and in which we are one of many universes—they would say, "Yes, what we really mean is, this is the place string theory seems to have led us, so we want to explore it." Is it necessarily the framework? I think most of them would say, "We don't know; we're just shooting in the dark, because this is our best approach to unifying general relativity and quantum mechanics, and we're going to explore where it leads us." They're not necessarily saying that this is definitely where it goes, because that's the nature necessarily of research: You don't know where it's going to lead, you just keep on going and see where it takes you.

STEINHARDT: But what angers people is even the idea that you might accept that possibility—that the ultimate theory has this googol of possibilities for the laws of physics. That should not be accepted. That should be regarded as an out-and-out failure, requiring some saving idea. The fact is that everywhere we look in the universe, we see only one set of laws. Also, the universe is smooth and uniform, smoother and more uniform than we need for humans to exist. Yet we are asked to accept the idea that the greater universe, beyond where we can see, is completely different. Is that science or is that metaphysics?

ISAACSON: That's exactly it. We were talking about why it is that it arouses such passion and then started directly debating string theory. I'd love to take it right back to Einstein. Twice you said something that I find very interesting, which is, we have to find a way to make his two grand pillars of 20th-century physics compatible, general relativity and quantum theory. Of course Einstein totally would believe that, because he loved unification, he loved unity. Secondly, he and Newton agreed on one big thing, which is that nature loves simplicity. But I've always wondered about the more metaphysical philosophical question: How do we know that God likes simplicity? How do we know he wants these things to be compatible? How do we know that quantum theory and relativity have to be reconcilable?

GREENE: There are actually some people who suggest that they don't necessarily have to be compatible. I've never really been convinced by their arguments at all. To me, it seems evident that the laws that we're talking about, and quantum theory in particular, are not meant just to describe small things—that's where it was developed and that's where its unusual features manifest themselves more strongly—but quantum theory is meant to be a theory that applies everywhere, on all scales.

General relativity starts as a theory that describes big things because that's where gravity matters, but when you look at the equations of general relativity, in principle they can apply on arbitrarily small scales. The thing is, when you get to really tiny scales, you notice that there's a deep incompatibility between the two theories, and moreover you realize that there are realms of the universe that enter those domains. You have, for instance, a black hole, which you can say begins as a star that then exhausts its nuclear fuel, collapses under its own weight, gets smaller and smaller—at some point the star gets so small that quantum mechanics really starts to matter in a significant way. Gravity matters the whole time, because it's so heavy. If those two theories don't work together, how do you describe what happens to this collapsing star?

ISAACSON: That's true of the Big Bang as well?

STEINHARDT: Yes, and that's why cosmology is the key battleground for trying to sort out how quantum physics and gravity relate. You can't avoid using them both to understand where the universe emerged from, or whether it had a beginning, or what happened before the Bang.

ISAACSON: And it's impossible to imagine a cosmos in which those two theories aren't in some fundamental way totally reconcilable?

STEINHARDT: It would be a mistake. It would be inconsistency.

GREENE: Although Freeman Dyson seems to have unusual views on this.

STEINHARDT: Yes, and I also would say I don't understand them—

ISAACSON: And that's where I got my question, but I tried to avoid Freeman Dyson because I was afraid I would totally misunderstand even his question.

STEINHARDT: But you ask a good question: "And how do we know?" The answer is, we don't know that things have to be simple. But a couple of interesting things have happened historically as we have followed that line of reasoning. We've managed to push the program of understanding the universe to small scales and large scales, by pursuing this approach of looking for simplicity. Particularly when we look at the cosmos. Now that we can see out to the farthest observable edges of the cosmos, we can see that the laws of physics are the same and that the physical conditions are also remarkably similar throughout the observable universe.

Until we have hard evidence to the contrary, I think we push this program of looking for unification and simplicity until it takes us as far as we can go. I have no sort of moral principles about the scientific method; I simply think it's the most efficient method humans have found yet of taking what we know and adding new knowledge. If we think up some other program of thinking that does better, we should adopt it, but at the moment—

ISAACSON: So it's a fundamental part of the program of thinking that the laws are unified at some level, and that we'll eventually get to more simplicity, not more Byzantine complexity, in the laws of nature?

GREENE: So long as you're willing to adjust the measure of simplicity and complexity as you learn more and more about the universe. If you were to present quantum mechanics to Newton, at first it might seem fairly complicated, because it uses a completely different body of mathematics, different kinds of ideas, invokes concepts that you can't directly see, and that certainly feels like it's a layer of complexity. But when you study it for many decades and you become used to its unusual features, you look at it and you see that it is just one simple little equation—Schrödinger's equation.

From a pure mathematical standpoint, it's a linear equation—in technical terms, the simplest kind of equation to analyze—and it describes data. So your sense of what's complicated and simple now gets shifted by a layer, because you're judging this framework, which to a 16th- or 17th-century scientist might seem really bizarre, but from a modern perspective it works and you attune your aesthetic sense so that it actually feels pretty elegant, and pretty simple.

ISAACSON: You use the word "elegant" often, which—

GREENE: It's become hackneyed, but if a theory is so simple that its deep equation can be put across a T-shirt in 20-point type, then we generally view that as fairly simple. Certainly that's the case with both general relativity and quantum mechanics.

ISAACSON: There's a wonderful book that Einstein wrote called *The Evolution of Physics* with Leopold Infeld, in 1938, which is not easy to find. I've gone over it again two or three times, because I just love the way it was written. It was written to make money for both of them, because it's the '30s, and Hitler, and refugees and stuff. It's a popular book, but it has a deep philosophical argument, and the publisher is reissuing the book because I was pushing them to get it out there.

The deep philosophical argument is that it will be a field theory approach that will work. It starts with Galileo. It talks about matter and particles and just makes the argument that in the end it's all going to be reconciled through field theory. It's about whether there's going to be a great distinction between a field theory and a theory of matter.

GREENE: You can even take that question one step further, which is: Is there even a distinction between, say, a field theory which has been so successful and a string theory, which appears

Brian Greene, Walter Isaacson, Paul Steinhardt

at first sight to invoke different ideas from what you'd get if you were just doing a purely field-theoretical approach?

One of the big ideas, one of the big results, in the last decade in string theory has been to find a close association, in fact an equivalence, between field theories and string theories. Even though string theory starts with a very different point of view and you can study it without it ever seeming to be a field theory, you realize—

ISAACSON: Isn't the mathematics different, though?

GREENE: The mathematics appears different at first, but one of the amazing kinds of discoveries in string theory in the last decade has been something called "dualities," which is, something can look one way and something else can look completely different, but if you study them with adequate intensity and adequate precision you can find that they're actually the same thing, just described in different languages. Like a book in French and in Sanskrit—they don't look the same, but if you have a dictionary that relates them, you can say, "Oh, this is the same book." Similarly, we have string theory framed in the language that's relevant to string theory—

ISAACSON: A mathematical language that's non-field theory—

GREENE: A mathematical language that feels non-field theory, that looks non-field theory. And then you have a field theory framed in the language of quantum field theory, and they seem different, they look different, but a lot of work has been done to set up the dictionary that allows you to say, "This element of this is that element of that," and vice versa, and you realize they're actually talking about the same theory, just in a different language.

ISAACSON: Is that sort of a reflection of the basic duality at the heart of quantum mechanics?

GREENE: No, this is a different kind of duality—and Paul has a different view of this—because the dualities that you're referring to sort of are inherent in each of these approaches, irrespective of the fact that they happen to be talking about the same theory. It's a completely unexpected and deep relationship between them, but fundamentally shows that string theory isn't that different from field theory; it's field theory—it's particular kinds of field theory—just organized in a different way, making it look different at first sight. But it basically confirms what you were saying from that book of seventy years ago—that field theory seems to be the tool that will take us to the next step.

STEINHARDT: I'm not sure how much faith I put in claims like that, because that's basically talking about what mathematics you use. It would be equivalent to saying the explanation is going to involve calculus. And while it's likely that field theory will be among the useful mathematical tools, it's likely that we're going to have to discover some new mathematics along the way to get to a final answer. I think most of us, Einstein included, tend to focus less on the tools and more on the underlying physical concepts.

I want to come back to one of the issues you raised—you asked about simplicity. I should have emphasized the following: It could very well have been that when we began to look out at the universe, we discovered different laws of physics, different gravitational forces, different electromagnetic forces, other

Brian Greene, Walter Isaacson, Paul Steinhardt

bizarre differences from what we see nearby—curiously, the very things that some people believe string theory predicts. Then we would be convinced experimentally that we don't live in a simple universe and that something like the stringy landscape picture is correct. We would also live in a universe which, due to its non-uniformity and our limited vantage point here on Earth, affords us no hope of understanding the universe in its entirety. Then the picture of a random universe would be compelling. The moment we begin to look into space, things would look so different from what we observe here—

ISAACSON: Einstein felt a little bit that way, I think, as quantum mechanics progressed in the late '40s; he was an old guy, but he kept discovering more and more particles and more and more forces that he was not even willing to accommodate, as he stood there at his blackboard with Valentine Bargmann and Peter Bergmann and all of his assistants. But he was vaguely depressed by the fact that nature seemed to like more and more forces and particles to be discovered that were not reconcilable.

STEINHARDT: Although I would guess that Einstein would love the concept of string theory. Not all string theorists feel the same way, but I view it as the fulfillment of Einstein's program of geometrizing the laws of physics. Einstein took gravity and turned it into wiggling Jell-O-like space, and now string theory turns everything in the universe, all forces, all constituents, into geometrical, vibrating, wiggling entities. String theory also uses the idea of higher dimensions, which is also something Einstein found appealing.

What I was commenting on earlier was where the string program has gone recently, which I described as a crash. I can't say for sure how Einstein would view it, but I strongly suspect he would reject the idea. I read an interesting quote from Einstein—I think in the '50s—in which he said that he had failed at constructing a unified theory and expressed his concern that it would be a very long time before there was any success. The reason, he claimed, which I thought was interesting, was because physicists no longer know about logic and philosophy. He didn't mention mathematics but, rather, logic and philosophy.

ISAACSON: What he really felt was that he had become more and more of a confirmed realist, or a scientific realist, and he felt that it was a philosophical question—that there was an underlying physical reality independent of our observations of it and that's what science was supposed to discover. And because that became so out of fashion with—he called them neo-positivists. You can call them whatever you want, but I assume most people in the forefront of quantum mechanics would not subscribe to the theory that there's underlying physical reality independent of any observations of it, which is a philosophical pillar on which you build science. And that was his philosophical problem.

STEINHARDT: I interpret it differently. By that time, something that might be called an American attitude toward physics had taken over. It was an attitude where the connection between physics and philosophy was broken. You were supposed to focus on what was calculable: Take your theory, make predictions with it, calculate with it, test if your calculations are right using experiments, and just stay away from philosophical questions. With this approach, the idea was that we can systematically inch our way forward in science.

ISAACSON: Yes, that's a good point. He could certainly have meant that, because he believed that, too.

STEINHARDT: So that in fact the reason a unified theory from that point of view would be beyond our reach is because if you didn't have deep philosophical principles to guide you, you just would never find your way. I thought that was an interesting quote, because it reflects some of my concerns about where both cosmology and fundamental physics are going—that maybe they've lost their way. Of course, it's unfashionable to appeal to philosophical viewpoints. But maybe that would be a healthy thing.

GREENE: Hey, we've got a research group with a philosopher as part of our group, so—

ISAACSON: But you've got art, you've got music—

GREENE: No, I'm not referring to those undertakings at all. We have a group that's trying to address questions such as the arrow of time. Where did the arrow of time come from? Why does time seem to unfold in one direction but not in reverse? This is a question that philosophers have studied intently for a long time, and they have refined the question in such a way that when we talk about various possible solutions, they're able to see the solutions and say, "Well, wait a second, you're actually assuming the answer in the solution in some hidden way that we've long since parsed out, and let me explain to you how that goes."

We've found it very useful to talk to philosophers who perhaps haven't studied quantum field theory with the kind of technical intensity a graduate student or a researcher in physics would have, but they've taken a step back and sort of looked at big questions and really thought them through at a fundamental level. And that's extraordinarily helpful.

ISAACSON: To get it back to the history: In the period from 1900 to 1915, which is to me—maybe because I'm a bit prejudiced by having worked on Einstein so much—a period of great explosive creativity. It's a period in which both quantum theory and relativity theory were developed. Much of that development was driven by philosophy and philosophers. If you ask Einstein who his most important influences were, he would get to Michelson and Morley at about number 3,500, if at all. Every now and then he would say, "Yes, I don't know if I ever read them. But Ernst Mach and David Hume—those are the people we were discussing all the time, and those are the people who led us to make the creative leaps we had to make in that period."

It's an interesting question to ask why it is that between 1900 and 1915 we have such creative leaps. Obviously in science, but even Stravinsky and Schoenberg saying, "OK, we don't have to stick to the classical bonds," or Proust and Joyce, or Picasso and Kandinsky—breaking the classical bonds. But especially in science, the leaps seemed to have been pushed by people like Ernst Mach, who are almost more philosophers than they are physicists.

STEINHARDT: My impression is that this began to break down with the developments of quantum mechanics in the 1920s, where after a certain effort to struggle with the interpretation of it, there did come this attitude that said, "OK, let's stop worrying about its interpretation; what we know we can do with it is to calculate and make predictions of new phenomena." And there were so many new phenomena to examine that it occupied generations of theorists to pursue this line of attack, ignoring interpretation and philosophy and just going forward in a straight line with these calculations. Once that historical connection with philosophy was broken, it became disparaged. It was considered that philosophy might even distract you from discovering something interesting. But now it might be that it's time to return to it.

You see, great progress had been occurring, because physicists were asking questions that could be tested almost immediately—the rapid interplay between experiment and theory was going back and forth for nearly a century. Every time a new observation would come in, or a new experiment would be performed, there would be a new question; you could do a new calculation; and someone else might do the next experiment. New physics was flowing from theory to experiment and back again very rapidly. But now we've reached the stage where the time between major experimental breakthroughs in fundamental physics is very long—decades, in the case of particle physics. We don't have experimental guidance, and we don't have the philosophical underpinnings, either. Maybe we don't just need new experiments. Maybe we need to look back to philosophy for guidance.

GREENE: In the PBS special on string theory that aired some time ago, there were a number of people who were interviewed about the fact that string theory had not yet made predictions that could be tested. And the framing of that fact by a number of physicists interviewed was, if string theory can't make, or doesn't make, those kinds of predictions, it becomes philosophy, not physics.

As I watched the series, I kept saying to myself, "the poor philosophers." Philosophy is not bad physics, it's not physics that hasn't reached its goal. It's just a way of analyzing pathways toward truth that perhaps don't use as much mathematics as the physicists and mathematicians typically do. There's a lot of insight yet to be tapped from the philosophical community, and I imagine we'll go through a cycle where that kind of interaction happens more and more.

STEINHARDT: Yes. In fact one of the interesting turns of events in string theory we've been talking about is the idea that there's a multiplicity of possibilities, and one of the approaches for dealing with it is the anthropic reasoning—to use the fact that we exist as a kind of selection principle. That turns out to be territory that philosophers have thought about quite a bit. They're far ahead of the physicists in terms of realizing the flaws and the trapdoors.

GREENE: I agree with you completely. Could we just come back to the assessment you gave of string theory a little while ago in terms of having crashed? That seems to me a pretty strong negative assessment, and I wonder if I'm hearing you fully, or if it's more nuanced.

When one looks at the history of string theory, the achievements have been manifold, as you are familiar with—the insights on spacetime singularities, mirror symmetry, topology change, the ability to understand certain symmetry structures, the ability to give insight into the possibility of having a generation structure in the families of matter particles, the insights that it's given on gauge theories as a general structure and in particular being able to realize gauge theories that we're familiar with—and all of these features, on top of its putting together general relativity and quantum mechanics. Now, I agree—and I'm actually all too happy to say—that we have a ways to go, because we've not made that direct contact with experimental observation. But to me, we have a road ahead of us that we still need to travel. Whether it will ultimately take us to those predictions or not, the future will tell. I don't think we can judge now and say the program has crashed. I can say the program has gone spectacularly far but we definitely have further to go until we know whether what we're doing is right or wrong. Is that not a good assessment in your view?

STEINHARDT: My view is more nuanced. This multiplicity, if that were to be the endpoint of the theory, is a crash. There's one of several possibilities—

GREENE: The only reason I interrupt you is because I've heard a number of people take a similar perspective, which is to listen to a couple of string theorists who are pushing one particular point of view: that maybe this is the endpoint of string theory— that there are many many universes, we're one of that many and there's no further explanation to be had. It may be right, it may be wrong, but I certainly don't at this point say that that's the endpoint of string theory. That's a way-station that some people are exploring, and others are pushing on other pathways.

STEINHARDT: You have a more reasonable attitude on this than others I've heard. I'd like it to be firmly recognized that googol possibilities is a crash, and that it's not an acceptable—

GREENE: But the many-universe version of string theory—is that what you mean?

STEINHARDT: Yes. But let me hasten to add that I can envision several ways to escape from this crash. One is to discover some new ideas in string theory showing that the multiplicity isn't really predicted by string theory. After all, the mathematical case is not firm. Or, second, even if there's a multiplicity, perhaps one can find some reason why almost everywhere in the universe should correspond to just one of these possibilities—namely, the one we actually observe. Or maybe string theory in its present form is fine but you have to change the cosmology, and that change removes the multiplicity. All these rescues are conceivable to me.

What I can't accept is the current view, which simply accepts the multiplicity. Not only is it a crash but it's a particularly nefarious kind of crash, because if you accept the idea of having a theory which allows an infinite number of possibilities, of which our observable universe is one, then there's really no way within science of disproving this idea. Whether a new observation or experiment comes out one way or the other, you can always claim afterward that we happen to live in a sector of the universe where that's so. In fact, this reasoning has already been applied recently, as theorists tried to explain the unexpected discovery of dark energy. The problem is that you can never disprove such a theory—nor can you prove it.

GREENE: You can imagine there are features that are consistent across all of these universes—

STEINHARDT: You could imagine it, but you could never prove it, experimentally.

GREENE: No, my point is, mathematically you could find that in each of these universes, property X always holds.

STEINHARDT: Do you mean, as derived from string theory? I don't believe that's true. I don't believe it's possible.

GREENE: Right, well, that's a belief—it's not based on any calculation

STEINHARDT: Well, I believe that if you came to me with such a theory, I could probably turn around within twenty-four hours and come up with an alternative theory in which property X wasn't universal after all. In fact, you almost know that's true from the conversation that's been happening in the field already, where someone says, "These properties are universal and these others are not." The next day, another theorist will write a paper saying, no, other properties are universal. There are simply no strong guidelines for deciding.

GREENE: I agree that that's definitely been the way things have unfolded. But I thought I heard you say that you couldn't imagine being able to disprove a theory that had this kind of framework, and I'm just setting up a way in which one could disprove it.

STEINHARDT: That may be true in principle, but in practice I don't think this would ever occur. If a version of string theory with a googolfold multiplicity of physical laws were to be disproved one day, I don't think proponents would give up on string theory. I suspect a clever theorist would come up with a variation that would evade the conflict. In fact, this has already been our experience with multiverse theories to date. In practice, there are never enough experiments or observations, or enough mathematical constraints, to rule out a multiverse of possibilities. By the same token, this means that there are no firm predictions that can definitively decide whether this multiplicity beyond our horizon is true or not.

GREENE: I agree with you. But just so I understand; you're saying that this one particular way in which one may think about string theory—for which the endpoint is many many universes—is unacceptable.

STEINHARDT: Right. I claim it needs to be fixed.

GREENE: But you also agree—just so it's clear—that that's not a crash in string theory per se; that's a particular way of approaching the theory that you would not advocate because the endpoint would be unacceptable. You need to go further—

STEINHARDT: That's right. So it's just what you were saying: Some people say that's the endpoint, and I'm saying that's not acceptable. If you believe that, it's time to abandon it.

GREENE: But it's those people who've crashed.

STEINHARDT: Yes, it's that point of view which is a crash and needs a fix. I'm not arguing that string theory should be abandoned. I think it holds too much promise. I'm arguing that it's in trouble and needs new ideas to save it.

But let's get back to Einstein. One interesting question to consider about Einstein is how his generation of physicists were radicals and were replaced by a generation of physicists that would be considered conservative.

ISAACSON: What's particularly interesting to me is that Einstein was a radical who in 1925 becomes replaced with Einstein who's a conservative. That's overstating it a bit, but right as he makes his last contribution of greatness to quantum theory—

basically the whole Bose-Einstein statistics—he almost instanta-neously is spinning around into a defensive crouch and resisting everything from the lack of realism to the lack of rigid causality in quantum mechanics, and he's calling them the young Turks at the Solvay conference 1927-1931, and they're calling him ridicu-lously conservative and saying he abandoned his radicalism, when he used to challenge everything.

It's a theme that goes well beyond physics, which is, Why is it that you used to think of yourself as a radical and then you become age fifty, and whether you're editing *Time* magazine or doing theoretical physics, you start saying things like, "No, we can't do that, we've tried that before and it doesn't work."

If I were to give a real reason for Einstein's basic conservatism, it would go back to what Paul was talking about, which is the philosophical, which is just that the concept of realism is so at the core. There are three or four reasons he doesn't like quantum mechanics, and if you had to pick one, it's not probabilities or the end of strict causality, even though he says strict causality is the greatest enduring gift that Newton gave us. It's the abandonment of realism, and to him that becomes a pillar of classical physics. If you have to define conservatism, I assume the definition would be defending the classical order as opposed to radically throwing out the old order. That's what he quits doing in 1925.

15
Einstein and Poincaré

Peter Galison

Joseph Pellegrino University Professor, director Collection of Historical Scientific Instruments, Harvard University; author, *Einstein's Clocks and Poincaré's Maps*

When the Einstein centenary was celebrated in 1979, the speakers at all of these great events spoke about physics only as theory. It seemed odd to me that somebody like Einstein, who had begun as a patent officer and who had been profoundly interested in experiments, had left such a thoroughly abstract image of himself. My interest in Einstein began in that period, but beyond Einstein I was intrigued by the startling way that experiment and theory worked together, fascinated by the abutting of craft knowledge hard against the great abstractions of theoretical physics.

For quite a number of years I have been guided in my work by the odd confrontation of abstract ideas and extremely concrete objects. Science history, sociology, and epistemology are for me very connected, and the kind of work that I do in the history of science is always propelled and illuminated through philosophical questions. For example, I'm interested in what counts as a demonstration. What does it mean to be done with a demonstration? How do experimenters distinguish between a real effect and artifacts of the apparatus or the environment? We think we know what it means to conclude a mathematical deduction, but what does it mean when I've shown something with a computer simulation? If I do a simulation and show that the tail of a comet forms into islands, have I demonstrated that, or is my result just the beginning of an explanation that then needs a more analytic mathematical derivation?

These are questions that even today puzzle across a myriad of fields. They are questions that are, inevitably both historical and epistemological—that is, they're about ordinary scientific practice and yet fundamentally philosophical. When I choose to work on a problem, it's usually because it is illuminated by these different beams of light, so to speak.

When I and a few other historians, sociologists, and philosophers began looking at instruments and laboratories back in the late 1970s, emphasizing experimentation in the history of science seemed rather odd. Most historians and philosophers were keen—in the aftermath of Thomas Kuhn's work—to show that all of science issued from theory. I suppose it was a kind of reaction against all those years of positivism, from the 1920s through the 1950s, when philosophers insisted that all knowledge came down to perception and observation. In any case, there wasn't really a body of serious work on what a laboratory was, where the lab came from, or how it functioned. Since then, inquiry into the history and dynamics of experimental practice has grown into a much larger domain of study. I'm interested not just in the laboratory itself, however, but also in the most abstract kinds of theories. Recently, for example, I've been writing about string theory—specifically, the confrontation between physicists and mathematicians as they try to sort out what ought to be a demonstration—in what is without doubt the most abstract form of science ever pursued.

But in every instance I'm above all intrigued by how philosophical questions illuminate and are illuminated by the practices of science, sometimes material, sometimes abstract. And I suppose I'm always interested in blasting away the mid-level generalizations, and exploring, as in *Einstein's Clocks, Poincaré's Maps*, the way the most abstract and the most concrete come together. Instead of thinking of a kind of smooth spectrum that goes from ultraviolet to infrared with everything in between, I'm interested in bending the edges of the spectrum to make the abstract and the concrete hit one another more directly.

When I began my work quite a number of years ago, the history of science was focused almost exclusively on the history of ideas and theories. Experiments and instruments, to the extent that they were of interest to anybody, were peripheral helpmates to the production of theory. I began by being interested in the way that certain kinds of instruments, or the way that instruments were used, shaped the way knowledge worked and the kinds of questions that people were asking. My first book, *How Experiments End*, was about how experimentalists decide they're looking at something real, whether it's using a small-scale table-top device or a huge experiment involving hundreds of people.

Then I turned to another subculture of physics, if you will, a subculture of people who are really interested in the machines themselves, not just in experimentation. I wanted to know how certain kinds of devices have carried a philosophy with them. For example, how did machines like cloud chambers and bubble chambers, which produce pictures, become the standard of evidence for a whole group of physicists across most of the 20th century? Or how did funny little objects like Geiger counters, which click when they're near something radioactive, produce a kind of statistical argument for new effects? What interested me was the contrast between the tradition of scientists who wanted to take pictures—who wanted to see in order to know—and another computing group who wanted to combine information more quantitatively—digitally, if you will—to produce a logic of demonstration. My second book, *Image and Logic*, is about these two huge, long-standing traditions within modern physics.

More recently I've been looking at what I consider to be the third subculture of physics: the theorists. I want to get at how theorists in the production of the most abstract ideas of physics— whether it's quantum field theory, relativity theory, or any other branch of theory—come to their concerns in relationship to very specific kinds of machines and devices in the world. Specifically, in *Einstein's Clocks, Poincaré's Maps* I pursue the vast concern about simultaneity in the late 19th century—what time was and what clocks were. This had a crucial dimension that was abstract and philosophical, but it also sprang from purely technological concerns. How, for example, do you make maps, or send signals across undersea cables? How do you coordinate and shunt trains so they don't smash into each other while going in opposite directions on the same track? Finally, my interest in theorists led me to look at the physics concerning the most pressing problem of the late 19th century, which was how electricity and magnetism work when an object moves through that all-pervasive entity people called "the ether."

My interest in the materiality of science goes back to my childhood. My great-grandfather, who lived into his mid-nineties, trained in Berlin and worked in Thomas Edison's laboratory as an electrical engineer, and I spent a great amount of time with him in his basement laboratory. I was completely riveted by what he did. It was the kind of laboratory you could imagine in a film about Dr. Frankenstein, with giant double-throw switches, arcs of electricity in the dark space, and bottles of mercury lining the shelves. I loved every bit of it. I left high school when I was seventeen, to study physics and mathematics at the Ecole Polytechnique in Paris for a year. I had a chance to learn from one of the great mathematicians, Laurent Schwartz. I'd been to France a fair amount, spoke French, and wanted to go there because I was very interested in European politics—these were wild times politically, toward the end of the Vietnam War. I thought the only chance I had of working in an interesting place would involve pursuing something in physics, so I wrote to various physics laboratories, and they must have taken me out of amusement at the idea of this American seventeen-year-old writing to the Polytechnique.

When I began, I was interested in philosophical questions and thought studying physics was a way to get at some of these problems. I worked in a laboratory on plasma physics, which is now done in gigantic machines in huge laboratories although at the time it was still possible to do small-scale experiments on devices not much bigger than a table. I became quite fascinated with the machinery, the signal generators, recording devices, oscilloscopes, and how theoretical knowledge about the world could come out of such material objects. In college at Harvard, I found a way, having done a fair amount of physics, to combine it with history and philosophy.

This brings me back to Einstein.

The Einstein we know today is mostly based on Einstein's later years, when he prided himself on his alienation from practically everything sociable and human, projecting an image of himself as a distracted, otherworldly character. We remember the Einstein who said that the best thing for a theoretical physicist would be to tend a lighthouse in quiet isolation from the world in order to be able to think pure thoughts. We have this picture of the theoretical physicist and project it backward to Einstein's miraculous year, 1905. It's easy enough to think of him as working a day job in a patent office merely to keep body and soul together while, in

Peter Galison

actuality, his real work was purely cerebral. Such a split existence never made sense to me; I wondered how his work in the details of machines and objects might connect to these abstract ideas, and began thinking about how relativity itself might have been lodged in the time, place, and machinery in which it was created.

Years later—one day in the summer of 1997—I was in a train station in northern Europe, looking down the platforms at these beautifully arranged clocks. The minute hands were all the same. I thought, "God, they made these extraordinary clocks back then! What an extraordinarily wonderful piece of machinery!" But I then noticed that the second hands were also all clicking along in sync. That meant the clocks were too good. So I thought that maybe they're not good clocks—maybe they're synchronized clocks bound together by electrical signals that advanced them together, in lockstep. Maybe Einstein had seen such clocks when he was writing his paper on relativity.

When I came back to the United States, I started poking around old Swiss, British, German, and American patents and industrial records, and it turns out that there was an enormous industry in coordinated clocks in the late 19th century. Suddenly the famous metaphor with which Einstein begins his 1905 paper began to look not so peculiar. Einstein asks us to interrogate what we mean by simultaneity. He says, imagine a train comes into a station where you are standing. If the hour hand of your watch just touches 7:00 as the train pulls in front of your nose, then you would say that the train's arrival and your watch showing 7:00 were simultaneous. But what does it mean to say that your clock ticks 7:00 at just the moment that a train arrives at a distant station?

Einstein goes on to develop a technique for saying what it would mean to coordinate clocks, and explains that this is what simultaneity is. This quasi-operational definition of simultaneity becomes the foundation of his theory and leads to his startling conclusions that simultaneity depends on frame of reference, that therefore length measurements are different in different frames of reference, and to all of the other famous and amazing results of relativity theory. Suddenly I could see that Einstein's seemingly abstract metaphor about trains and stations was actually both entirely metaphorical and yet altogether literal. Far from being the only person worried about the meaning of simultaneity—a lighthouse keeper in splendid isolation—there was a vast industry of people worrying about what it meant to say that a train was arriving at a distant train station. And they were determining simultaneity by sending electrical signals down telegraph lines to distant stations in ways very much like the way Einstein was describing in that fateful paper.

So I began to look further, wondering who else would have been worrying about simultaneity in the late 19th century. It turns out that the great French philosopher, mathematician, and physicist Henri Poincaré, had much the same idea as Einstein. He also wanted to criticize the idea of absolute simultaneity and to make it something that could be measured.

Instead of trains and stations, Poincaré chose for his key metaphor the exchange of a telegrapher's signal down a line. In his famous philosophical article of January 1898, Poincaré says that simultaneity is really just the exchange of signals, like two telegraphers trying to determine how much longitudinal difference there is between them. You see, if the Earth were stationary, we could find our longitude simply by looking up to see which stars were straight above us. But the Earth turns, so to compare two longitudes—that is, the stars above two different sites—you have to make the measurement at the same time. Consequently, for

Peter Galison

centuries map makers have worried about simultaneity and how to determine it. By the late 19th century, people were exchanging electrical time signals across the oceans via undersea cables, and what is interesting is that Poincaré was right in the middle of it. In 1899 he was elected president of the Bureau of Longitude in Paris. Then, in December 1900, he brought his new definition of time from philosophy and technology into the heartland of physics. He showed that if the telegraphers coordinated their clocks when moving through the ether, their clocks would "appear" to be simultaneous even though from the "true" ether-rest system they were not. But now the new definition of simultaneity stood central for Poincaré in all three domains: philosophy, technology, and physics.

Though Poincaré was as famous as any mathematician or philosopher of his time, he was also a man of enormous engineering skills, trained as a sophisticated engineer at the Polytechnique and Ecole des Mines in Paris and later becoming one of the Polytechnique's most illustrious professors. It is Poincaré's situatedness that intrigues me: Like Einstein, when Poincaré invoked the longitude-finding telegraphers, he was speaking both metaphorically and literally. He was changing a central concept for all physics and at the same time addressing the real practices of map makers.

Though less well known by far than Einstein, at the turn of the century Poincaré's popular philosophical books, *Science and Hypothesis* and *Science and Values*, were best-sellers in France. They had a profound effect on the modern philosophy of science and today are still read in philosophy courses. They were also translated into many other languages very early on, including German and English, and widely distributed. He opened up whole new areas of mathematics, including the mathematics of topology. He helped invent the science of chaos, and all that we understand of the science of complexity owes an enormous amount to him. He contributed enormously to what became relativity theory and is important in many other branches of physics. He was truly a polymath and went on to do things in engineering. He was one of the people who rescued the Eiffel Tower from being taken down after the International Exhibition for which it was built, because he saw a way of using it as a military antenna. In fact, in large measure under Poincaré's direction, the Eiffel Tower itself became an enormous antenna that would send time signals all over the world, allowing longitude finders from Canada to the tip of Africa to do their work. Moving back and forth smoothly between high engineering and abstract mathematics, he left an enormous legacy across many fields, always reasoning concretely, visually—as an abstract engineer, so to speak. His thoughts on time were no exception.

After learning more about Poincaré, I tried to understand how he and Einstein could have radically reformulated our ideas of time and space by looking at the way that philosophically abstract concerns, physics concerns, and these technological problems of keeping trains from bashing into each other and coordinating map making across the empires might fit into a single story. It begins with an extraordinarily simple idea: that two events are simultaneous if I can make clocks at the two events say the same thing. How do I coordinate these clocks? I send a signal from one to the other and take into account the time it takes for the signal to get there. That's the basic idea, but all of relativity theory, $E = mc^2$, and so much of what Einstein does follows from it. The question is, Where did this idea come from? Albert Einstein and Henri Poincaré were the two people who worked out this practical, almost operational idea of simultaneity, and I want to

Peter Galison

see them as occupying points of intersection of technological, philosophical, and physical reasoning. They were the two people who stood dead center in those triple crossing points.

Sometimes people ask me, "What is really at the base of Einstein's and Poincaré's account of simultaneity? Is it really physics, or fundamentally technology, or does it come down to philosophy?" I think those are wrong ways of putting the question. That is to say, to me it's like asking if the Place de l'Etoile is truly in the Avenue Foch or the Avenue Victor Hugo. The Place de l'Etoile is a place because it's at the intersection of those great avenues. And that's what happens here. We're looking at an extraordinary moment when philosophy, physics, and technology cross, precisely because of the intersection of three very powerful streams of action and reasoning at the turn of the century. It's like having a triple spotlight focused on one position in an enormous theater; it's triply illuminated. It was important to railroad engineers and map makers that they knew how to define simultaneity. It was important to philosophers to figure out what time is, what a clock is, and how to think about what defined time—mechanical clocks or astronomical phenomena or some sort of abstract time that lay behind all appearances. And it was important to physicists to understand what simultaneity was in order to know how to interpret the most important equations of physics, Maxwell's equations concerning electricity and magnetism. Poincaré and Einstein were the two people—more than anyone else—who were concerned with all three parts of that intersection, and that's why they need to be understood together. Of course, clocks did not cause relativity, any more than relativity caused the transformation of modern clock synchronization.

In human terms, Einstein and Poincaré are fascinating because in some ways you couldn't imagine two people closer. They had common friends, published in many of the same places, and were working intensively on many of the same problems. They were both at the top of their professions, both enjoyed writing for broader audiences, both were taken very seriously by philosophers, and both had serious technological-engineering interests and training. Yet they couldn't have been farther apart. In a certain way they remind me of Freud, for whom it was almost unbearable to read Nietzsche, because—as Freud said on several occasions—Nietzsche's ideas were too close and yet organized around a different approach.

Poincaré and Einstein, who had two of the largest scientific correspondences of the 19th and 20th centuries, including thousands of letters to and from other people, never exchanged a single postcard over the entirety of their overlapping lives. They met once, toward the end of Poincaré's life, when Poincaré presided over a session at a vitally important physics conference where Einstein was talking about his new ideas about the quantum of light. At the end of this session, Poincaré said that Einstein's presentation was so different from what physics should be—namely, that it could be represented with causal interactions, with good differential equations, with clear presentations of principles and consequences—that he simply found it unbearable, and ended by making it clear that what Einstein was saying was so contradictory that anything could follow from it. It was a disaster for science, he thought. Einstein, for his part, went home and scribbled a note to a friend in which he recounted the wonderful work that had been done by various colleagues, how much he admired, even loved, the physicist Hendrik Lorentz, but disparaged Poincaré, who simply seemed to understand nothing. They passed like ships in the night, each, on relativity, unable to acknowledge the other's existence. Yet a few weeks after their ill-starred meeting,

Poincaré wrote a letter of recommendation for Einstein for a job that was very important to him. It was a stunning letter that said, essentially, that this young man may well be up to some of the greatest things and even if only a few of his wild ideas turn out to be true he's a person of extraordinary importance. It was a letter of enormous grace and generosity. They never directly exchanged another word and never met again.

The contrast between Einstein and Poincaré, and their different understandings of what they were doing represent two grand competing visions of modern science for the 20th century. Although the equations that Poincaré and Einstein come up with around relativity theory are very similar—essentially identical—Poincaré always thought of what he was doing as fixing, repairing, or continuing the past by applying reason to it. As one of his relatives once put it, he was filling in the blank spots on the map of the world. Einstein was willing to do things rather differently, to say that the old way of proceeding is too complicated, too filled with piecemeal solutions, and that what we need is something that starts over again with the classical purity of a few stark principles. Poincaré saw himself in some ways as saving an empire—the empire of France, no doubt, but also the empire of 19th-century physics. His was a grand ambition, but it's a different kind of modernism from Einstein's. It's a reparative, ameliorative modernism, a modernism with all the rational hopefulness of a Third Republic Frenchman. Einstein's is a much more disruptive, classifying, purifying modernism. It is only by understanding this triple intersection of philosophy, physics, and technology that one can really grasp what each of these alternative visions of the new century is about.

You might ask, and I've often wondered, how to think about this kind of event in the present. That is to say, Is there an analogy now to this kind of triple intersection? Here's how I think about it: When you consider Poincaré and Einstein, you're dealing with an attempt to understand time-coordination and the synchronization of clocks at a huge variety of scales. In some ways, they're trying to figure out how to coordinate clocks inside a single room, or observatory, or a block, or a whole city, at the same time that the people who are worried about these things are also sending cables across the Pacific and Atlantic Oceans. Einstein and Poincaré are not just worrying about such planetary scales but also about how to coordinate clocks in different reference frames in the universe as a whole. They're asking, "What does synchronization mean? What does simultaneity mean?" These are questions that occur at every scale, from the smallest to the largest, from philosophy and physics all the way down to electrical wiring along train tracks. In that sense, it's unlike most questions we ask in science, since it doesn't have the character of starting out as something purely abstract that then becomes applied physics and engineering, eventually ending up on the factory floor. It's also not a Platonic ascension, or a naïve version of Marxism, in which machines and machine-shop relations are slowly abstracted to ever wider spheres until they become a theory of the universe.

Questions of the conventionality of time, of how it becomes equated with physical processes and procedures, are key to all of the domains considered. And the metaphor we need for this back-and-forth between the practical and the philosophical is not just one of condensation from the abstract to the concrete. Nor is it one of evaporation, in which water grows less dense as it passes into a vapor. Instead, more helpful is that phenomenon physicists call critical opalescence. Ordinary opalescence is that oyster-shell color in which you see all colors reflected, that remarkable surface of a pearl or the inner surface of certain shells in which you can see red, green, and white, all at once. Critical opalescence in matter occurs under very particular circumstances—for example, in a system of water and vapor held under just the right combination of temperature. At this critical point, something quite extraordinary happens. The liquid starts to evaporate and condense at every scale, from a couple of molecules to a whole system. Suddenly, because droplets form of every size—from a couple of molecules coming together to the whole of the system—light of every wavelength reflects back. If you shine in blue, you see blue; if you shine in red, you see red; and if you shine in yellow, you see yellow.

That's the kind of metaphor we need to look at a situation like this. Poincaré and Einstein are flipping back and forth between philosophical questions, physics questions, and practical questions. At the end of the 1890s, Poincaré was publishing in journals for map makers and longitude finders at the same time he was publishing in physics journals and in the *Journal of Metaphysics and Morals*. In his thinking he was flipping back and forth extraordinarily quickly between these three domains of philosophy, physics, and technology.

Now, one can ask how this might compare to the present. What kind of critical opalescence marks the science of recent times? It seems to me fairly rare, but one place you might see it is in the collection of sciences that have grown up around computation. Here, ideas about the mind, about how computers function, and about science, codes, and mathematical physics all come together. Von Neumann thinks about the mind and its organs (memory, input–output, processing) as a way of designing a programmed computer. The programmed computer then becomes a model for the mind. The ideas of information, which are encoded into the development of computation, also become ways to understand language and communication more generally, and again feed back into devices. Information, entropy, and computation become metaphors for us at a much broader level. Such opalescent moments are not that common, surely rarer than whatever it is that we mean by scientific revolutions. They're something else. No, points of critical opalescence in this sense point to science in times and places where we're starting to think with and through machines at radically different scales, where we are flipping back and forth between abstraction and concreteness so intensively that they illuminate each other in fundamentally novel ways—in ways not captured by models of simple evaporation or condensation. When we see such opalescence, we should dig into them, and deeply, for they are transformative moments of our cultures.

Peter Galison

16
Thinking About the Universe on the Larger Scales

Raphael Bousso

Professor of theoretical physics, Berkeley

INTRODUCTION by Leonard Susskind

The parable of the blind men and the elephant is a perfect metaphor for the universe and for the physicists who try to grasp its larger shape. Each man feels a part of the elephant and tries to visualize its overall essence: "It's a wall"; "It's a rope"; "It's a tree." They almost come to blows. The universe, even more than the elephant, is too big for any one perspective, and most of us are busy squabbling about some small part.

Fortunately, now and then someone comes along who's brave enough, bold enough, and with clear enough vision, to have a chance of seeing the bigger picture. Raphael Bousso is one of those few.

Thinking About the Universe on the Larger Scales

We can ask ourselves questions at different levels of generality—or profundity, if you will—and I guess as a scientist it's important to strike a balance. We tend to not make much progress if we decide to work on the deepest, most far-reaching questions straight out. It's good to have those as a compass, but it's important to break things up in some way, and the way that I would break things up is to say, "The far-reaching questions are things like how do we unify all the laws of nature, how do you do quantum gravity, how do you understand how gravitation and quantum mechanics fit together, how does that fit in with all the other matter and forces that we know?" Those are really far-reaching and important questions.

Another far-reaching question is, "What does the universe look like on the largest scales?" How special is the part of the universe that we see? Are there other possibilities? Those questions are connected with each other, but in order to try to answer them we have to come up with specific models, with specific ways to think about these questions, with ways to break them down into pieces—and of course, most important, with ways to relate them to observation and experiment.

One important hint that came along on the theoretical side a long time ago was string theory, which wasn't invented for this sort of deep-sounding questions. It was invented to understand something about the strong force, but then it took on its own life and became this amazing structure that could be explored and that started spitting out these answers to questions you hadn't even thought of asking yet, such as quantum gravity. It started doing quantum gravity for you.

This is a controversial issue. There are other approaches to this problem of quantum gravity. I personally find the string theory by far the most well developed and the most promising, and so I find myself looking for hints about the answers to these kinds of questions that are outlined by using string theory, by exploring the properties of this theory, by asking it what it tells us about these questions.

Another hint that helps us break things up and lower the questions down to accessible levels is, of course, observational: What do we see when we look out the window? The one thing that's really remarkable that we see, and it's remarkable in the way that

Raphael Bousso

the question of why the sky is not bright at night is remarkable, is . . . It sounds stupid, but when you really think about it, that's a profound question and it needs an explanation: "Why isn't there a star everywhere you look?" A similar kind of question is, "Why is the universe so large?" It's actually extremely remarkable that the universe is so large, from the viewpoint of fundamental physics. A lot of amazing things have to happen for the universe not to be incredibly small, and I can go into that. One of the things that has to happen is that the energy of empty space has to be very, very small for the universe to be large. And in fact, just by looking out the window and seeing that you can see a few miles out, it's an experiment that already tells you that the energy of empty space is a ridiculously small number, 0.000 and then dozens of zeros and then a 1. Just by looking out the window, you learn that.

The funny thing is that when you calculate what the energy of empty space should be using theories you have available—really well-tested stuff that's been tested in accelerators, like particle theory, the standard model, things that we know work—you use that to estimate the energy of empty space, and you can't calculate it exactly on the dot. But you can calculate what the size of different contributions is, and they're absolutely huge. They should be much larger than what you already know it can possibly be—again, not just by factor of 10 or 100 but by a factor of billions, of billions of billions of billions.

This requires an explanation. It's only one of the things that has to go right in order for the universe to become as large as we see it, but it's one of the most mysterious properties that turned out to be right for the universe to become large. It needs an explanation.

Funnily enough, because we knew that that number had to be so small—that is, the energy of empty space, the weight of empty space, had to be so small—it became the lore within at least a large part of the physics community that it was probably zero, for some unknown reason. And one day we'd wake up and discover why it's exactly zero. But instead, one day in 1998 we woke up and discovered that it's non-zero. One day we woke up in 1998 and discovered that cosmologists had done some experiments that looked at how fast the universe has been accelerating at different stages of its life, and discovered that the universe had started to accelerate its expansion—whereas we used to think that what it would do is explode at the Big Bang and then kind of get slower and slower in the way that galaxies expand away from one another. Instead, it's like you went off the brakes and stepped on the gas pedal a few billion years ago. The universe is accelerating. That's exactly what a universe does if the energy of empty space is non-zero and positive, and you could look at how fast its acceleration is happening and deduce the actual value of this number. In the last thirteen years a lot of independent observations have come together to corroborate this conclusion.

It's still true that the main thing we needed to explain is why the cosmological constant, or the energy of empty space, isn't huge. But now we also know that the explanation was definitely not going to be that for some symmetry reason that number is exactly zero. And so we needed an explanation that would tell us why that number is not huge but also not exactly zero.

The amazing thing is that string theory, which wasn't invented for this purpose, managed to provide such an explanation, and in my mind this is the first serious contact between observation, experiment on the one side, and string theory on the other. It was always interesting to have a consistent theory of quantum gravity. It's very hard to write that down in the first place, but it

Raphael Bousso

turned out that string theory has exactly the kind of ingredients that make it possible to explain why the energy of empty space has this bizarre, very small, but non-zero value.

I thought I was going to become a mathematician and then decided to study physics instead, at the last minute, because I realized that I actually cared about understanding nature, and not just some abstract—perhaps beautiful, but abstract—construct. I went to Cambridge, the one in England, for my PhD. I worked with Stephen Hawking on questions of quantum properties of black holes and how they might interplay with early-universe cosmology. I went on to Stanford for a postdoc. At Stanford, I talked a lot to Andrei Linde and Lenny Susskind, and we all felt it was time for string theory to have some sort of say about cosmology—that string theory had not really taught us enough about cosmology—and we started thinking in various different ways about how string theory might do that.

One idea that was floating around was called the holographic principle. This is an idea that deals with the question of how much information you need to describe a region of spacetime at the most fundamental level, and surprisingly the answer is not infinity. Even more surprising, the answer doesn't grow with the volume of the region. As 't Hooft and Susskind had first intuited, the answer is related to the area surrounding the region. But the idea didn't really fully work, especially in cosmology. So one of the topics I worked on was trying to understand whether this idea of the holographic principle is really correct, whether it can be formulated in such a way that it makes sense in all imaginable spacetime regions, in cosmology, inside black holes, and not just in some harmless place where gravity is not important. That turned out to be true, and so that was very exciting.

Another topic I started thinking about was trying to understand the small but non-zero value of the cosmological constant, the energy of empty space, or, as people like to call it, dark energy. I worked on that subject with Joe Polchinski, at KITP, in Santa Barbara, and we realized that string theory offers a way of understanding this, and I would argue that that is the leading explanation currently of this mysterious problem. From Stanford, I went on to a postdoc at Santa Barbara, and then a number of small stops here and there, including one year at Harvard. In 2004, I joined the faculty at Berkeley, where I am now a professor in the Physics Department.

I don't do experiments, in the sense that I would walk into a lab and start connecting wires to something. But it matters tremendously to me that the theory I work on is supposed to actually explain something about nature. The problem is that the more highly developed physics becomes, the more we start asking questions which for technological reasons are not in the realm of day-to-day experimental feedback. We can't ask about quantum gravity and expect at the same time to be getting some analogue of the spectroscopic data that in the late 19th century fed the quest for quantum mechanics. And I think it's a perfectly reasonable reaction to say, "Well, in that case I think the subject is too risky to work on." But I think it's also a reasonable reaction to say, "Well, but the question, it's obviously a sensible one." It's clearly important to understand how to reconcile quantum mechanics and general relativity. They're both great theories, but they totally contradict each other, and there are many reasons to believe that by understanding both we will learn profound things about how nature works. Now, it could be that we're not smart enough to do this—in particular, without constant feedback from experiments—but we have been pessimistic at so many junctures in the past and we found a way around.

I don't think we're going to understand a lot about quantum gravity by building more particle accelerators. We'll understand a lot of other things, even a few things about quantum gravity, but rather indirectly. But we'll look elsewhere, we'll look at cosmological experiments, we'll use the universe to tell us about very high energies. We'll come up with ideas that we can't even dream about right now. I'm always in awe of the inventiveness of my experimental colleagues, and I don't doubt that they will deliver for us eventually.

It has been said that it's been a golden age for cosmology in the last fifteen years or so, and it's true. I was very lucky with timing. When I was a graduate student, the COBE satellite was launched, and started flying and returning data, and that really marked the beginning of an era where cosmology was no longer the sort of subject where there were maybe one or two numbers to measure and people had uncertainties on, say, how fast the universe expands. They couldn't even agree on how fast the galaxies are moving away from one another. And from this, we move to a data-rich age where you have unbelievably detailed information about how matter is distributed in the universe, how fast the universe is—not just expanding right now but the expansion history, how fast it was expanding at earlier times, and so on. Things were measured that seemed out of reach just a few years earlier, and so indeed it's no longer possible to look down on cosmology as this sort of hand-waving subject where you can say almost anything and never be in conflict with the data. In fact, a lot of theories have gone down the road of being eliminated by data in the past fifteen years or so, and several more are probably going to go down that road pretty soon.

An example of a theory that has been ruled out is one of the ideas for how structure originally formed in the universe. Why isn't the universe just some sort of homogeneous soup? Why are there clumps of galaxies here, empty spaces there, another galaxy here? How did that come about, and how did the particular way in which these objects are distributed come about? Why are they the size they are, why aren't they larger or smaller, why isn't there maybe just one galaxy which is really huge, and the rest, all we can see, empty? This clearly needs an explanation.

There were a number of different theories on the market. One of them was inflation. One of the nice things about it was that it was not originally invented for the purpose of explaining this, but it turned out to have something to say about this question. Then there are other theories that were also reasonably well motivated, such as cosmic strings—not the same thing as the string-theory strings but objects that we call topological defects. Basically, these are objects which are stringlike, and energy is sort of locked into them in a way that it can't get out, because of the way that the universe cooled down as it was expanding very early on. And cosmic strings would lead to some sort of structure if you have the right kind of cosmic strings around, but it makes very different detailed predictions about what that structure looks like, what kind of imprints it leaves in the cosmic microwave background that satellites like COBE have now measured so well, and that *Planck* is currently measuring with incredible precision.

We already know enough about the cosmic microwave background that we can completely rule out the possibility that cosmic strings are responsible for structure formation. It's, of course, possible that there are cosmic strings out there, but they would have to be of a type that has not had any impact on structure formation.

Inflation looks really good. It's not like we have a smoking-gun confirmation of it, but it has passed so many tests—it could have been ruled out quite a few times by now—that it, I would say, is looking really interesting right now.

Inflation comes in many detailed varieties, but it does make a number of rather generic predictions, and unless you work very hard to avoid them, they come with pretty much every inflation model you grab off the shelf. One of those predictions is that the spatial geometry of the universe would be flat. It should be the kind of geometry you learn about in high school, as opposed to the weird kind of geometry that mathematicians study in university, and that has turned out to be the case. To within a percent precision, we now know that the universe is spatially flat. Inflation predicts a particular pattern of perturbations in the sky, and again, to the extent that we have the data—and we have very precise data by now—there was plenty of opportunity to rule out that prediction, but inflation still stands. So there are a number of general predictions that inflation makes which have held up very well, but we're not yet at a point where we can say it's this particular make and model of inflation that's the right one and not this other one. We're zooming in. Some types of inflation have been ruled out, large classes of models have been ruled out, but we haven't zoomed in on the one right answer, and that might still take a while, I would expect.

I was saying that string theory has in a way surprised us by being able to solve a problem that other theories, including some that were invented for that purpose alone, had not been able to address—i.e., the problem of why empty space weighs so little, why there's so little dark energy. The way string theory does this is similar to the way we can explain the enormous variety of what we see when we look at the chair, the table, and the sofa in this room. What are these things?

They're basically a few basic ingredients—electrons, quarks, and photons. You've got five different particles, and you put them together, and now you've got lots and lots of these particle. There are very few fundamental ingredients, but you have many copies of them. You have many quarks, you have many electrons, and when you put them together you have a huge number of possibilities of what you can make. It's just like with a big box of Legos: There are lots of different things you can build out of that. With a big box of quarks and electrons, you can build a table if you want, or you can build a chair if you want. It's your choice. And strictly speaking, if I take one atom and I move it over here to a slightly different place on this chair, I've built a different object. These objects in technical lingo will be called solutions of a certain theory called the standard model. If I have a block of iron, I move an atom over there, it's a different solution of the standard model.

The fact that there are innumerably many different solutions of the standard model does not, of course, mean that the standard model of particle physics—this triumph of human thinking—is somehow unbelievably complicated, or that it's a theory of anything, or that it has no predictive power. It just means that it's rich enough to accommodate the rich phenomenology we actually see in nature, while at the same time starting from a very simple setup. There are only certain quarks. There is only one kind of electron. There are only certain ways you can put them together, and you cannot make arbitrary materials with them. There are statistical laws that govern how very large numbers of atoms behave, so even though things look like they get incredibly complicated, they start simplifying again when you get to really large numbers.

In string theory, we're doing a different kind of building of iron blocks. String theory is a theory that wants to live in ten

Raphael Bousso

dimensions—nine spatial dimensions and one of time. We live in three spatial dimensions and one of time, or at least so it seems to us. And this used to be viewed as a little bit of an embarrassment for string theory—not fatal, because it's actually fairly easy to imagine how some of those spatial dimensions could be curled up into circles so small that they wouldn't be visible even under our best microscopes. But it might have seemed nicer if the theory had just matched up directly with observation.

It matches up with observation very nicely when you start realizing that there are many different ways to curl up the six unwanted dimensions. How do you curl them up? Well, it's not like they just bend themselves into some random shape. They get shaped into a small bunch of circles, whatever shape they want to take, depending on what matter there is around. Similar to how the shape of your Lego car depends on how you put the pieces together, or the shape of this chair depends on how you put the atoms in it together, the shape of the extra dimensions depends on how you put certain fundamental string-theory objects together.

Now, string theory actually is even more rigorous about what kind of fundamental ingredients it allows you to play with than the Lego company or the standard model. It allows you to play with fluxes, D-branes, and strings. And these are objects that we didn't put into the theory; the theory gives them to us and says, "This is what you get to play with." But depending on how it warps strings and D-branes and fluxes in the extra six dimensions, these six dimensions take on a different shape. In effect, this means that there are many different ways of making a three-dimensional world. Just as there are many ways of building a block of iron or a Lego car, there are many different ways of making a three-plus-one dimensional-seeming world.

Of course, none of these worlds are truly three-plus-one dimensional. If you could build a strong enough accelerator, you could see all these extra dimensions. If you could build an even better accelerator, you might be able to even manipulate them and make a different three-plus-one dimensional world in your lab. But naturally you would expect that this happens at energy scales that are currently, and probably for a long time, inaccessible to us. But you have to take into account the fact that string theory has this enormous richness in how many different three-plus-one dimensional worlds it can make.

Joe Polchinski and I did an estimate, and we figured that there should be not millions or billions of different ways of making a three-plus-one dimensional world, but ten to the hundreds, maybe 10^{500} different ways of doing this. This is interesting for a number of reasons, but the reason that seemed the most important to us is that it implies that string theory can help us understand why the energy of the vacuum is so small. Because, after all, what we call "the vacuum" is simply a particular three-plus-one dimensional world—what that one looks like when it's empty. And what that one looks like when it's empty is basically, it still has all the effects from all this stuff you have in the extra dimensions, all these choices you have there about what to put.

For every three-plus-one dimensional world, you expect that in particular the energy of the vacuum is going to be different—the amount of dark energy, or cosmological constant, is going to be different. And so you have 10^{500} ways of making a three-plus-one dimensional world, and in some of them, just by accident, the energy of the vacuum is going to be incredibly close to zero.

The other thing that's going to happen is that in about half of these three-plus-one dimensional worlds, the vacuum is going to have positive energy. So even if you don't start out the universe

Raphael Bousso

in the right one—where by "right one" I mean the one that later develops beings like us to observe it—you could start it out in a pretty much random state, another way of making a three-dimensional world. What would happen is that it would grow very fast, because positive vacuum energy needs acceleration, as we observed today in the sky. It will grow very fast and then by quantum mechanical processes it would decay, and you would see changes in the way matter is put into these extra dimensions, and locally you would have different three-plus-one dimensional worlds appearing.

This is not something I made up. This is actually an effect which predates string theory, which goes back to calculations by Sidney Coleman and others in the '70s and '80s, and which doesn't rely on any fancy gravity stuff. This is actually fairly pedestrian physics, which is hard to really argue with. What happens is, the universe gets very, very large; all these different vacua, three-dimensional worlds that have positive weight, grow unboundedly and decay locally; and new vacua appear that try to eat them up but don't eat them up fast enough. So the parents grow faster than the children can eat them up, and so you make everything. You fill the universe with these different vacua, these different kinds of regions in which empty space can have all sorts of different weights. Then you can ask, "Well, in such a theory, where are the observers going to be?" To just give the most primitive answer to this question, it's useful to remember the story about the holographic principle that I told you a little bit earlier.

If you have a lot of vacuum energy, then even though the universe globally grows and grows and grows, if you sit somewhere and look around, there's a horizon around you. The region that's causally connected—where particles can interact and form structure—is inversely related to the amount of vacuum energy you have. This is why I said earlier that just by looking out the window and seeing that the universe is large, we know that there has to be very little vacuum energy. If there's a lot of vacuum energy, the universe is a tiny little box from the viewpoint of anybody sitting in it. The holographic principle tells you that the amount of information in the tiny little box is proportional to the area of its surface. If the vacuum energy has this sort of typical value that it has in most of the vacua, that surface allows for only a few bits of information. So whatever you think observers look like, they probably are a little bit more complicated than a few bits.

And so you can immediately understand that you don't expect observers to exist in the typical regions. They will exist in places where the vacuum energy happens to be unusually small due to accidental cancellations between different ingredients in these extra dimensions, and where, therefore, there's room for a lot of complexity. And so you have a way of understanding both the existence of regions in the universe somewhere with very small vacuum energy, and also of understanding why we live in those particular, rather atypical regions.

What's interesting about this is the idea that maybe the universe is a very large multiverse with different kinds of vacua in it; this was actually thrown around independently of string theory for some time, in the context of trying to solve this famous cosmological-constant problem. But it's not actually that easy to get it all right. If you just imagine that the vacuum energy got smaller and smaller and smaller as the universe went on, that the vacua are nicely lined up, with each one you decay into having slightly smaller vacuum energy than the previous one, you cannot solve this problem. You can make the vacuum energy small, but you also empty out the universe. You won't have any matter in it.

What was remarkable was that string theory was the first theory that provided a way of solving this problem without leading to a prediction that the universe is empty—which is obviously fatal and immediately rules out that approach. That, to me, was really remarkable, because the theory is so much more rigid; you don't get to play with the ingredients, and yet it was the one that found a way around this impasse and solved this problem.

I think the things that haven't hit Oprah yet, and which are up and coming, are questions like, "Well, if the universe is really accelerating its expansion, then we know that it's going to get infinitely large, and that things will happen over and over and over." And if you have infinitely many tries at something, then every possible outcome is going to happen infinitely many times, no matter how unlikely it is. This is something that predates this string-theory multiverse I was talking about. It's a very robust question, in the sense that even if you believe string theory is a bunch of crap, you still have to worry about this problem, because it's based on just observation. You see that the universe is currently expanding in an accelerated way, and unless there's some kind of conspiracy that's going to make this go away very quickly, it means that you have to address this problem of infinities. But the problem becomes even more important in the context of the string landscape, because it's very difficult to make any predictions in the landscape if you don't tame those infinities.

Why? Because you want to say that seeing this thing in your experiment is more likely than that thing, so that if you see the unlikely thing you can rule out your theory, the way we always like to do physics. But if *both* things happen infinitely many times, then on what basis are you going to say that one is more likely than the other? You need to get rid of these infinities. This is called, at least among cosmologists, the measure problem. It's probably a really bad name for it, but it stuck.

That's where a lot of the action is right now. That's where a lot of the technical work is happening, that's where people are, I think, making progress. I think we're ready for Oprah, almost, and I think that's a question where we're going to come full circle, we're going to learn something about the really deep questions, about what the universe is like on the largest scales, how quantum gravity works in cosmology. I don't think we can fully solve this measure problem without getting to those questions, but at the same time, the measure problem allows us a very specific way in. It's a very concrete problem. If you have a proposal, you can test it, you can rule it out, or you can keep testing it if it still works, and by looking at what works, by looking at what doesn't conflict with observation, by looking at what makes predictions that seem to be in agreement with what we see, we're actually learning something about the structure of quantum gravity. So I think that it's currently a very fruitful direction. It's a hard problem, because you don't have a lot to go by. It's not like it's an incremental, tiny little step. Conceptually it's a very new and difficult problem. But at the same time it's not that hard to state, and it's remarkably difficult to come up with simple guesses for how to solve it that you can't immediately rule out. And so we're at least in the lucky situation that there's a pretty fast filter. You don't have a lot of proposals out there that have even a chance of working.

The thing that's really amazing, at least to me, is in the beginning we all came from different directions at this problem, we all had our different prejudices. Andrei Linde had some ideas, Alan Guth had some ideas, Alex Vilenkin had some ideas. I thought I was coming in with this radically new idea that we shouldn't think of the universe as existing on this global scale that no one observer can actually see—that it's actually important to think about what can happen in the causally connected region to one observer. What can you do, in any experiment, that doesn't actually conflict with the laws of physics and require superluminal propagation. We have to ask questions in a way that conforms to the laws of physics if we want to get sensible answers.

I thought, "OK, I'm going to try this." This is completely different from what these other guys are doing, and it's motivated by the holographic principle that I talked about earlier. I was getting pretty excited, because this proposal didn't run into immediate catastrophic problems like a lot of other simple proposals did. When you go into the details, it spit out answers that were really in much better agreement with the data than what we had had previously from other proposals. And I still thought that I was being original.

But then we discovered—and actually my student I-Sheng Yang played a big role in this discovery—that there's a duality, an equivalence of sorts, and a very precise one, between this global way of looking at the universe that most cosmologists had favored and what we thought was our radical new local causal connected way of thinking about it. In a particular way of slicing up the universe in this global picture in a way that's, again, motivated by a different aspect of the holographic principle, we found that we kept getting answers that looked exactly identical to what we were getting from our causal-patch proposal. For a while we thought, "OK, this is some sort of approximate, accidental equivalence," and if we asked detailed enough questions we were going to see a difference. Instead, what we discovered was a proof of equivalence—that these two things are exactly the same way of calculating probabilities, even though they're based on what mentally seemed like totally different ways of thinking about the universe.

That doesn't mean we're on the right track. Both of these proposals could be wrong. Just because they're equivalent doesn't mean they're right. But a lot of things have now happened that didn't have to happen. A lot of things have happened that give us some confidence that we're on to something, and at the same time we're learning something about how to think about the universe on the larger scales.

Raphael Bousso

17
Quantum Monkeys

Seth Lloyd

Quantum mechanical engineer, MIT; author, *Programming the Universe*

It's no secret that we're in the middle of an information-processing revolution. Electronic and optical methods of storing, processing, and communicating information have advanced exponentially over the last half-century. In the case of computational power, this rapid advance is known as Moore's Law. In the 1960s, Gordon Moore, the ex-president of Intel, pointed out that the components of computers were halving in size every year or two, and consequently the power of computers was doubling at the same rate. Moore's law has continued to hold to the present day. As a result, these machines that we make, these human artifacts, are on the verge of becoming more powerful than human beings themselves, in terms of raw information-processing power. If you count the elementary computational events that occur in the brain and in the computer—bits flipping, synapses firing—the computer is likely to overtake the brain in terms of bits flipped per second in the next couple of decades.

We shouldn't be too concerned, though. For computers to become smarter than us is not really a hardware problem; it's more a software issue. Software evolves much more slowly than hardware, and indeed much current software seems designed to junk up the beautiful machines we build. The situation is like the Cambrian explosion—a rapid increase in the power of hardware.

Who's smarter, humans or computers? is a question that will get sorted out some million years hence—maybe sooner. My guess would be that it will take hundreds or thousands of years until we actually get software we could reasonably regard as useful and sophisticated. At the same time, we're going to have computing machines that are much more powerful quite soon.

Most of what I do in my everyday life is to work at the very edge of this information-processing revolution. Much of what I say to you today comes from my experience in building quantum computers—computers where you store bits of information on individual atoms. About ten years ago, I came up with the first method for physically constructing a computer in which every quantum—every atom, electron, and photon—inside a system stores and processes information. Over the last ten years, I've been lucky enough to work with some of the world's great experimental physicists and quantum mechanical engineers to actually build such devices. A lot of what I'm going to tell you today is informed by my experiences in making these quantum computers. During this meeting, Craig Venter claimed that we're all so theoretical here that we've never seen actual data. I take that personally, because most of what I do on a day-to-day basis is try to coax little superconducting circuits to give up their secrets.

The digital information-processing revolution is only the most recent revolution, and it's by no means the greatest one. For instance, the invention of movable type and printing has had a much greater impact on human society so far than the electronic revolution. There have been many information-processing revolutions. One of my favorites is the invention of the so-called Arabic—actually Babylonian—numbers, in particular, zero. This amazing invention, very useful in terms of processing and registering information, came from the ancient Babylonians and then

Seth Lloyd

moved to India. It came to us through the Arabs, which is why we call it the Arabic number system. The invention of zero allows us to write the number 10 as "one, zero." This apparently tiny step is in fact an incredible invention that has given rise to all sorts of mathematics, including the bits—the binary digits—of the digital computing revolution.

Another information-processing revolution is the invention of written language. It's hard to argue that written language is not an information-processing revolution of the first magnitude.

Another of my favorites is the first sexual revolution—that is, the discovery of sex by a living organism. One of the problems with life is that if you don't have sex, then the primary means of evolution is via mutation. Almost 99.9 percent of mutations are bad. Being from a mechanical engineering department, I would say that when you evolve only by mutation, you have an engineering conflict: Your mechanism for evolution happens to have all sorts of negative effects. In particular, the two prerequisites for life—evolve, but maintain the integrity of the genome—collide. This is what's called a coupled design, and that's bad. However, if you have sexual selection, then you can combine genes from different genomes and get lots of variation without, in principle, ever having to have a mutation. Of course, you still have mutations, but you get a huge amount of variation for free.

I wrote a paper a few years ago that compared the evolutionary power of human beings to that of bacteria. The point of comparison was the number of bits per second of new genetic combinations that a population of human beings generated, compared with the number generated by a culture of bacteria. A culture of bacteria in a swimming pool of seawater has about a trillion bacteria, reproducing once every thirty minutes. Compare this with the genetic power of a small town with a few thousand people in New England—say, Peyton Place—reproducing every thirty years. Despite the huge difference in population, Peyton Place can generate as many new genetic combinations as the culture of bacteria a billion times more numerous. This assumes that the bacteria are only generating new combinations via mutation, which of course they don't, but for this purpose we will not discuss bacteria having sex. In daytime TV, the sexual recombination and selection happens much faster, of course.

Sexual reproduction is a great revolution. Then, of course, there's the grandmother or granddaddy of all information-processing revolutions, life itself. The discovery, however it came about, that information can be stored and processed genetically and that this could be used to encode functions inside an organism that can reproduce is an incredible revolution. It happened 4 to 5 billion years ago on Earth—maybe earlier, if one believes that life developed elsewhere and then was transported here. At any rate, since the universe is only 13.8 billion years old, it happened sometime in the last 13.8 billion years.

We forgot to talk about the human brain—or should I say, my brain forgot to talk about the brain? There are many information-processing revolutions, and I'm presumably leaving out many thousands we don't even know about but which were equally as important as the ones we've discussed.

To pull a Kuhnian maneuver, the main thing I'd like to point out about these information-processing revolutions is that each one arises out of the technology of the previous one. Electronic information-processing, for instance, comes out of the notion of written language—of having zeroes and ones, the idea that you can make machines to copy and transmit information. A printing press is not so useful without written language. Without spoken language, you wouldn't come up with written language. It's hard to speak if you don't have a brain. And what are brains for but to help you have sex? You can't have sex without life. Music came from the ability to make sound, and the ability to make sound evolved for the purpose of having sex. You either need vocal cords to sing with or sticks to beat on a drum with. To make sound, you need a physical object. Every information-processing revolution requires either living systems, electromechanical systems, or mechanical systems. For every information-processing revolution, there is a technology.

OK, so life is the big one, the mother of all information-processing revolutions. But what revolution occurred that allowed life to exist? I would claim that, in fact, all information-processing revolutions have their origin in the intrinsic computational nature of the universe. The first information-processing revolution was the Big Bang. Information-processing revolutions come into existence because at some level the universe is constructed of information. It is made out of bits.

Of course, the universe is also made out of elementary particles, unknown dark energy, and lots of other things. I'm not advocating that we junk our normal picture of the universe as being constructed out of quarks, electrons, and protons. But in fact it's been known ever since the latter part of the 19th century that every elementary particle, every photon, every electron, registers a certain number of bits of information. Whenever two elementary particles bounce off each other, those bits flip. The universe computes.

The notion that the universe is, at bottom, processing information sounds like some radical idea. In fact, it's an old discovery, dating back to Maxwell, Boltzmann, and Gibbs, the physicists who developed statistical mechanics from 1860 to 1900. They showed that in fact the universe is fundamentally about information. They, of course, called this information "entropy," but if you look at their scientific discoveries through the lens of 20th-century technology, what in fact they discovered was that entropy is the number of bits of information registered by atoms. So in fact it's scientifically uncontroversial that the universe at bottom is processing information. My claim is that this intrinsic ability of the universe to register and process information is actually responsible for all the subsequent information-processing revolutions.

How do we think of information these days? The contemporary scientific view of information is based on the theories of Claude Shannon. When Shannon came up with his fundamental formula for information, he went to the physicist and polymath John von Neumann and said, "What shall I call this?" and von Neumann said, "You'll call it H, because that's what Boltzmann called it," referring to Boltzmann's famous H-theorem. The founders of information theory were well aware that the formulas they were using had been developed back in the 19th century to describe the motions of atoms. When Shannon talked about the number of bits in a signal that can be sent down a communications channel, he was using the same formulas to describe it that Maxwell and Boltzmann had used to describe the amount of information, or the entropy, required to describe the positions and momenta of a set of interacting particles in a gas.

What is a bit of information? Let's get down to the question of what information is. When you buy a computer, you ask how many bits its memory can register. A bit comes from a distinction between two different possibilities. In a computer, a bit is a little electric switch, which can be open or closed; or it's a capacitor that can be charged, which is called 1, or uncharged, which is called 0. Anything that has two distinct states registers a bit of information. At the elementary-particle level, a proton can have two distinct states: spin-up or spin-down. Each proton registers one bit of information. In fact, the proton registers a bit whether it wants to or not, or whether this information is interpreted or not. It registers a bit merely by the fact of existing. A proton possesses two different states and so registers a bit.

We exploit the intrinsic information-processing ability of atoms when building quantum computers, because many of our quantum computers consist of arrays of protons interacting with their neighbors, each of which stores a bit. Each proton would be storing a bit of information whether we were asking them to flip those bits or not. Similarly, if you have a bunch of atoms zipping around, they bounce off each other. Take two helium atoms in a child's balloon. The atoms come together and they bounce off each other and then they move apart again. Maxwell and Boltzmann realized that there's essentially a string of bits that attach to each of these atoms to describe its position and momentum. When the atoms bounce off each other, the string of bits changes, because the atoms' momenta change. When the atoms collide, their bits flip.

The number of bits registered by each atom is well known and has been quantified ever since Maxwell and Boltzmann. Each particle—for instance, each of the molecules in this room— registers something on the order of 30 or 40 bits of information as it bounces around. This feature of the universe—that it registers and processes information at its most fundamental level—is scientifically uncontroversial, in the sense that it has been known for 120 years and is the accepted dogma of physics.

The universe computes. My claim is that this intrinsic information-processing ability of the universe is responsible for the remainder of the information-processing revolutions we see around us, from life up to electronic computers. Let me repeat the claim: It's a scientific fact that the universe is a big computer. More technically, the universe is a gigantic information processor capable of universal computation. That's the definition of a computer.

If he were here, Marvin Minsky would say, "Ed Fredkin and Konrad Zuse, back in the 1960s, claimed that the universe was a computer, a giant cellular automaton." Konrad Zuse was the first person to build an electronic digital computer, around 1940. He and Ed Fredkin, at MIT, came up with this idea that the universe might be a gigantic type of computer called a cellular automaton. This is an idea that has since been developed by Stephen Wolfram. The idea that the universe is some kind of digital computer is, in fact, an old claim as well.

Thus, my claim that the universe computes is an old one dating back at least half a century. This claim could actually be substantiated from a scientific perspective. One could prove, by looking at the basic laws of physics, that the universe is or is not a computer, and if so, what kind of computer it is. We have very good experimental evidence that the laws of physics support computation. I own a computer, and it obeys the laws of physics, whatever those laws are. We know the universe supports computation, at least on a macroscopic scale. My claim is that the universe supports computation at its most tiny scale. We know that the universe processes information at this level, and we know that at the larger level it's capable of doing universal computations and creating things like human beings.

The thesis that the universe is, at bottom, a computer, is in fact an old notion. The work of Maxwell, Boltzmann, and Gibbs established the basic computational framework more than a century ago. But for some reason, the consequences of the computational nature of the universe have yet to be explored in a systematic way. What does it mean to us that the universe computes? This question is worthy of significant scientific investigation. Most of my work investigates the scientific consequences of the computational universe.

One of the primary consequences of the computational nature of the universe is that the complexity we see around us arises in a natural way, without outside intervention. Indeed, if the universe computes, complex systems like life must necessarily arise. So describing the universe in terms of how it processes information, rather than describing it solely in terms of the interactions of elementary particles, is not some kind of empty exercise. Rather, the computational nature of the universe has dramatic consequences.

Let's be more explicit about why something that's computationally capable, like the universe, must necessarily spontaneously generate the kind of complexity that's around us. There's a famous story, "Inflexible Logic," by Russell Maloney, which appeared in *The New Yorker* in 1940, in which a wealthy dilettante hears the phrase that if you had enough monkeys typing, they would type the works of Shakespeare. Because he's got a lot of money, he assembles a team of monkeys and a professional trainer and he has them start typing. At a cocktail party, he has an argument with a Yale mathematician who says that this is really implausible, because any calculation of the odds of this happening will show it will never happen. The gentleman invites the mathematician up to his estate in Greenwich, Connecticut, and he takes him to where the monkeys have just started to write out *Tom Sawyer* and *Love's Labour's Lost*. They're doing it without any single mistake. The mathematician is so upset that he kills all the monkeys. I'm not sure what the moral of this story is.

The image of monkeys typing on typewriters is quite old. I spent a fair amount of time this summer going over the Internet and talking with various experts around the world about the origins of this story. Some people ascribe it to Thomas Huxley in his debate with Bishop Wilberforce in 1858, after the appearance of *The Origin of Species*. From eyewitness reports of that debate, it's clear that Wilberforce asked Huxley from which side of his family, his mother's or his father's, he was descended from an ape. Huxley said, "I would rather be descended from a humble ape than from a great gentleman who uses considerable intellectual gifts in the service of falsehood." A woman in the audience fainted when he said that. They didn't have R-rated movies back then.

Although Huxley made a stirring defense of Darwin's theory of natural selection during this debate, and although he did refer to monkeys, apparently he did not talk about monkeys typing on typewriters, because for one thing typewriters as we know them had barely been invented in 1859. The erroneous attribution of the image of typing monkeys to Huxley seems to have arisen because Arthur Eddington, in 1928, speculated about monkeys typing all the books in the British Library. Subsequently Sir James Jeans ascribed the typing monkeys to Huxley.

In fact, it seems to have been the French mathematician Emile Borel who came up with the image of typing monkeys, in 1907. Borel was the person who developed the modern mathematical theory of combinatorics. Borel imagined a million monkeys each typing ten characters a second at random. He pointed out that these monkeys could in fact produce all the books in all the richest libraries of the world. He then went on to dismiss the probability of them doing so as infinitesimally small.

It is true that the monkeys would, in fact, type gibberish. If you plug "monkeys typing" into Google, you'll find a website that will enlist your computer to emulate typing monkeys. The site lists records of how many monkey years it takes to type out the opening bits of various Shakespeare plays and the current record is 17 characters of *Love's Labour's Lost* over 483 billion monkey years. Monkeys typing on typewriters generate random gobbledygook.

Before Borel, Boltzmann advanced a monkeys-typing explanation for why the universe is complex. The universe, he said, is just a big thermal fluctuation. Like the flips of a coin, the universe is, in fact, just random information. His colleagues soon dissuaded him from this position, because it's obviously not so. If it were, then every new bit of information you got that you hadn't received before would be random. But when our telescopes look out in space, they get new information all the time and it's not random. Far from it: The new information they gather is full of structure. Why is that?

To see why the universe is full of complex structure, imagine that the monkeys are typing into a computer rather than a typewriter. The computer, in turn, rather than just running Microsoft Word, interprets what the monkeys type as an instruction in some suitable computer language, like Java. Now, even though the monkeys are still typing gobbledygook, something remarkable happens. The computer starts to generate complex structures.

At first this seems odd: Garbage in, garbage out. But in fact there are short, random-looking computer programs that will produce very complicated structures. For example, one short, random-looking program will make the computer start proving all provable mathematical theorems. A second short, random-looking program will make the computer evaluate the consequences of the laws of physics. There are computer programs to do many things, and you don't need a lot of extra information to produce all sorts of complex phenomena from monkeys typing into a computer.

There's a mathematical theory called algorithmic information which can be thought of as the theory of what happens when monkeys type into computers. This theory was developed in the early 1960s by Ray Solomonoff in Cambridge, Mass.; Gregory Chaitin, who was then a fifteen-year-old *enfant terrible* at IBM in Brazil, and then Andrey Kolmogorov, who was a famous Russian academic mathematician. Algorithmic information theory tells you the probability of producing complex patterns from randomly programmed computers. The bottom line is that if monkeys start typing into computers, there's a very high probability that they'll produce things like the laws of chemistry, autocatalytic sets, or prebiotic kinds of life. Monkeys typing into computers make up a reasonable explanation for why we have complexity in our universe. Monkeys typing into a computer have a reasonable probability of producing almost any computable form of order that exists. You would not be surprised, in this monkey universe, to see all sorts of interesting things arising. You might not get *Hamlet,* because something like *Hamlet* requires huge sophistication and the evolution of societies, etc. But things like the laws of chemistry, or autocatalytic sets, or some kind of prebiotic form of protolife are the kinds of things you would expect to see happen.

To apply this explanation to the origin of complexity in our universe, we need two things: a computer and monkeys. We have the computer, which is the universe itself. As was pointed out a century ago, the universe registers and processes information systematically at its most fundamental level. The machinery is there to be typed on. So all you need is monkeys. Where do you get the monkeys?

Seth Lloyd

The monkeys that program our universe are supplied by the laws of quantum mechanics. Quantum mechanics is inherently chancy. You may have heard Einstein's phrase, "God does not play dice." Einstein was wrong. God does play dice. In the case of quantum mechanics, Einstein was, famously, wrong. In fact, it is just when God plays dice that these little quantum blips, or fluctuations, get programmed into our universe.

For example, Alan Guth has done work on how such quantum fluctuations form the seeds for the formation of large-scale structure in the universe. Why is our galaxy here rather than somewhere 100 million light-years away? It's here because way back in the very, very, very, very early universe there was a little quantum fluctuation that made a slight overdensity of matter somewhere near here. This overdensity of matter was very tiny, but it was enough to make a seed around which other matter could clump. The structure we see, like the large-scale structure of the universe, is in fact made by quantum monkeys typing.

We have all the ingredients, then, for a reasonable explanation of why the universe is complex. You don't require very complicated dynamics for the universe to compute. The computational dynamics of the universe can be very simple. Almost anything will work. The universe computes. Then the universe is filled with little quantum monkeys, in the form of quantum fluctuations, that program it. Quantum fluctuations get processed by the intrinsic computational power of the universes and eventually give rise to the order that we see around us.

18
The Nobel Prize and After

Frank Wilczek

Herman Feshbach Professor of Physics, MIT; recipient, 2004
Nobel Prize in physics; author, *The Lightness of Being*

In retrospect, I realize now that having the Nobel Prize hovering out there but never quite arriving was a heavy psychological weight; it bore me down. It was a tremendous relief to get it. Fortunately, it turns out I didn't anticipate that getting it is fantastic fun—the whole bit. There are marvelous ceremonies in Sweden, it's a grand party, and it continues and is still continuing. I've been going to big events several times a month.

The most profound aspect of it, though, is that I've really felt from my colleagues something I didn't anticipate: an outpouring of genuine affection. It's not too strong to call it love. Not for me personally, but because our field, theoretical fundamental physics, gets recognition and attention. People appreciate what's been accomplished, and it comes across as recognition for an entire community and an attitude toward life that produced success. So I've been in a happy mood.

But that was a while ago, and the ceremonial business gets old after a while, and takes time. Such an abrupt change of life encourages thinking about the next stage. I was pleased when I developed a kind of three-point plan that gives me direction. Now I ask myself, when I'm doing something in my work: Is it relating to point one? Is it relating to point two? Is it relating to point three? If it's not relating to any of those, then I'm wasting my time.

Point one is in a sense the most straightforward. An undignified way to put it would be to say it's defending turf, or pissing on trees, but I won't say that: I'll say it's following up ideas in physics that I've had in the past that are reaching fruition. There are several I'm very excited about now. The great machine at CERN, the LHC, is going to start operating in about a year. Ideas—about unification and supersymmetry and producing Higgs particles—that I had a big hand in developing twenty to thirty years ago are finally going to be tested. Of course, if they're correct that'll be a major advance in our understanding of the world and very gratifying to me personally.

Then there's the area of exotic behavior of electrons at low temperature, so-called anyons, which is a little more tech. It was thought for a long time that all particles were either bosons or fermions. In the early '80s, I realized there were other possibilities, and it turns out that there are materials in which these other possibilities can be realized—where the electrons organize themselves into collective states that have different properties from individual electrons and actually do obey the peculiar new rules, and are anyons. This is leading to qualitatively new possibilities for electronics. I call it anyonics. Recently, advanced anyonics has been notionally bootstrapped into a strategy for building quantum computers that might even turn out to be successful.

In any case, whether it's successful or not, the vision of anyonics—this new form of electronics—has inspired a lot of funding, and experimentalists are getting into the game. Here, similarly, there are kinds of experiments that have been in my head for twenty years but are very difficult and people needed motivation and money to do them, that are now going to be done. It's a lot of fun to be involved in something that might actually have practical consequences and might even change the world. This stuff also, in a way, brings me back to my childhood, because when I was growing up my father was an electrical engineer and was taking home circuit diagrams, and I really admired these things. Now I get to think about making fundamentally new kinds of circuits, and it's very cool. I really like the mixture of abstract and concrete.

At a deeper level, what excites me about quantum computing and this whole subject of quantum information-processing is that it touches such fundamental questions that potentially it could lead to qualitatively new kinds of intelligences. It's notorious that human beings have a hard time understanding quantum mechanics; it's hard for humans to relate to its basic notions of superpositions of states—that you can have Schrödinger's cat that's both dead and alive—that are not in our experience. But an intelligence based on quantum computers—mechanical quantum thinking—from the start would have that in its bones, so to speak, or in its circuits. That would be its primary way of thinking. It's quite challenging but fascinating to try to put yourself in the other guy's shoes, when that guy has a fundamentally different kind of mind, a quantum mind.

It's almost an embarrassment of riches, but some of the ideas I had about axions turn out to go together very very well with inflationary cosmology—and to get new pictures for what the dark matter might be. It ties into questions about anthropic reasoning, because with axions you get really different amounts of dark matter in different parts of the multiverse. The amounts of dark matter would be different elsewhere, and the only way to argue about how much dark matter there should be turns out to be that if you have too much dark matter, life as we know it couldn't arise.

Frank Wilczek

There's a lot of stuff in physics that I really feel I have to keep track of and do justice to. That's point one.

The second point is another way of having fun: looking for outlets, cultivating a public, not just thinking about science all the time. I'm in the midst of writing a mystery novel that combines physics with music, philosophy, sex, the rule that only three people at most can share a Nobel Prize—and murder. Or was it suicide? When a four-person MIT-Harvard collaboration makes a great discovery in physics—they figure out what the dark matter is—somebody's got to go. That project, and I hope subsequent projects, will be outlets in reaching out to the public and bringing in all of life and just having fun.

The third point is what I like to call the Odysseus project. I'm a great fan of Odysseus, the wanderer, who had adventures and was very clever. I really want to do more great work—not following up what I did before but doing essentially different things. I got into theoretical physics almost by accident; when I was an undergraduate, I had intended to study how minds work and neurobiology. But it became clear to me rather quickly, at Chicago, that that subject at that time wasn't ripe for the kind of mathematical analytical approach that I really like and get excited about and am good at. I switched and majored in mathematics and eventually wound up in physics.

But I've always maintained that interest, and in the meantime the tools available for addressing those questions have improved exponentially. Both in terms of studying the brain itself—imaging techniques and genetic techniques and a variety of others—but also the inspiring model of computation. The explosion of computational ability and understanding of computer science and networks is a rich source of metaphors and possible ways of thinking about the nature of intelligence and how the brain works. That's a direction I really want to explore more deeply. I've been reading a lot; I don't know exactly what I want to do, but I have been nosing out what's possible and what's available. I think it's a capital mistake, as Sherlock Holmes said, to start theorizing before you have the data. So I'm gathering the data.

Quantum computing is an inspiring vision, but at present it's not clear what the technical means to carry it off are. There are a variety of proposals. It's not clear which is the best, or if any of them are practical.

Let me backtrack a little bit, though, because even before you get to a full-scale quantum computer, there are information-processing tasks for which quantum mechanics could be useful with much less than a full-scale quantum computer. A full-scale quantum computer is extremely demanding: You have to build various kinds of gates, you have to connect them in complicated ways, you have to do error correction—it's very complicated. That's sort of like envisioning a supersonic aircraft when you're at the stage of the Wright brothers. However, there are applications that I think are almost in hand.

The most mature is for a kind of cryptography: You can exploit the fact that quantum mechanics has this phenomenon that's roughly called "collapse of the wave function"—I don't like that, I don't think that's a really good way to talk about it, but for better or worse, that's the standard terminology. Which in this case means that if you send a message that's essentially quantum mechanical—in terms of the direction of spins of photons, for instance—then you can send photons one by one with different spins and encode information that way. If someone eavesdrops on this, you can tell, because the act of observation necessarily disturbs the information you're sending. So that's very useful.

Frank Wilczek

If you want to transmit messages and make sure they haven't been eavesdropped, you can have that guaranteed by the laws of physics. If somebody eavesdrops, you'll be able to tell. You can't prevent it, necessarily, but you can tell. If you do things right, the probability of anyone being able to eavesdrop successfully can be made negligibly small. So that's a valuable application that's almost tangible. People are beginning to try to commercialize that kind of idea.

I think in the long run the killer application of quantum computers will be doing quantum mechanics. Doing chemistry by numbers, designing molecules, designing materials by calculation. A capable quantum computer would let chemists and materials scientists work at a another level, because instead of having to mix up the stuff and watch what happens, you can just compute. We know exactly what the equations are that govern the behavior of nuclei and electrons and the things that make up atoms and molecules.

So in principle, it's a solved problem: To figure out chemistry, just compute. We don't know all the laws of physics, but it's essentially certain that we know the adequate laws of physics with sufficient accuracy to design molecules and predict their properties with confidence. But our practical ability to solve the equations is limited. The equations live in big multidimensional spaces, and they have a complicated structure, and, to make a long story short, we can't solve any but very simple problems. With a quantum computer we'll be able to do much better.

As I sort of alluded to earlier, it's not decided yet what the best long-term strategy is for achieving powerful quantum computers. People are doing simulations and building little prototypes. There are different strategies being pursued based on nuclear spins, electron spins, trapped atoms, anyons.

I am very fond of anyons, because I worked at the beginning on the fundamental physics involved. It was thought, until the late '70s and early '80s, that all fundamental particles, or all quantum mechanical objects that you could regard as discrete entities, fell into two classes: so-called bosons, after the Indian physicist Satyendra Nath Bose, and fermions, after Enrico Fermi. Bosons are particles such that if you take one around another, the quantum mechanical wave function doesn't change. Fermions are particles such that if you take one around another the quantum mechanical wave function is multiplied by a minus sign. It was thought for a long time that those were the only consistent possibilities for the behavior of quantum mechanical entities. In the late '70s and early '80s, we realized that in two-plus-one dimensions—not in our everyday three-dimensional space (plus one dimension for time), but in planar systems—there are other possibilities. In such systems, if you take one particle around another, you might get not a factor of 1 or -1 but multiplication by a complex number. There are more general possibilities.

More recently, the idea that when you move one particle around another it's possible not only that the wave function gets multiplied by a number but that it actually gets distorted and moves around in bigger space—that idea has generated a lot of excitement. Then you have this fantastic mapping from motion in real space, as you wind things around each other, to motion of the wave function in Hilbert space—in quantum mechanical space. It's that ability to navigate your way through Hilbert space that connects to quantum computing and gives you access to a gigantic space with potentially huge bandwidth that you can play around with in highly parallel ways, if you're clever about the things you do in real space.

But in anyons we're really at the primitive stage. There's very

Frank Wilczek

little doubt that the theory is correct, but the experiments are at a fairly primitive stage—they're just breaking now.

Quantum mechanics is so profound that it genuinely changes the laws of logic. In classical logic, a statement is either true or false; there's no real sense of in-between. But in quantum mechanics you can have statements or propositions encoded in wave functions that have different components, some of which are true, some of which are false. When you measure, the result is indeterminate. You don't know what you're going to get. You have states, meaningful states of computation—what you can think of as states of consciousness—that simultaneously contain contradictory ideas and can work with them simultaneously. I find that concept tremendously liberating and mind-expanding. The classic structures of logic are really far from adequate to do justice to what we find in the physical world.

To do justice to the possible states—the possible conditions that just a few objects can be in; say, five spins—classically you would think you would have to say for each one whether it's up or down. At any one time, they're in some particular configuration. In quantum mechanics, every single configuration—there are thirty-two of them, up or down for each spin—has some probability of existing. So simultaneously to do justice to the physical situation, instead of just saying that there's some configuration these objects are in, you have to specify roughly that there's a certain probability for each one and those probabilities evolve. But that verbal description is too rough, because what's involved is not probabilities; it's something called amplitudes. The difference is profound. Whereas probabilities have a kind of independence, with amplitudes the different configurations can interact with one other. There are different states in the physical reality, and they're interacting with each other. Classically they would be different things that couldn't happen simultaneously; in quantum theory, they coexist and interact with one another.

That also goes to this issue of logic I mentioned before. One way of representing true or false that's famously used in computers is, you have true as 1 and false as 0; spin-up is true, spin-down is false. In quantum theory, the true statement and the false statement can interact with each other, and you can do useful computations by having simultaneous propositions that contradict each other, sort of interacting with each other, working in creative tension. I just love that idea. I love the idea of opposites coexisting and working with one another. Come to think of it, it's kind of sexy.

Realizing this vision will be a vast enterprise. It's hard to know how long it's going to take to get something useful, let alone something competitive with the kind of computing we already have developed, which is already very powerful and keeps improving—let alone create new minds that are different from and more powerful than the kind of minds we're familiar with. We'll need to progress on several fronts.

You can set aside the question of engineering, if you like, and ask: "Suppose I had a big quantum computer, what would I do with it, how would I program it, what kind of tasks could it accomplish?" That's a mathematical investigation. You abstract the physical realization away. Then it becomes a question for mathematicians, and even philosophers have got involved in it.

Then there's the other big question: "How do I build it?" How do I build it in practice? That's a question very much for physicists. In fact there's no winning design yet; people have struggled to make even very small prototypes. My intuition, though, is that when there's a really good idea, progress could be very rapid. That's what I'm hoping for and going after. I have glimmers of how it might be done, based on anyons.

I've been thinking about this sort of thing on and off for a long time. I pioneered some of the physics, but other theorists, including Alexei Kitaev and my former student Chetan Nayak, have taken things to another level. There's now a whole field called topological quantum computing, with its own literature and conferences, and it's moving fast. What has changed is that now a lot of people—and, in particular, experimentalists—have taken it up.

The most exciting thing that can happen in the career of a theoretical physicist is when theoretical dreams that started as fantasies, as desires, become projects that people work hard to build. There is nothing like it; it's the ultimate tribute. At one moment you have just a glimmer of a thought and at another moment squiggles on paper. Then one day you walk into a laboratory and there are all these pipes, and liquid helium is flowing, and currents are coming in and out with complicated wiring, and somehow all this activity supposedly corresponds to those little thoughts you had. When this happens, it's magic.

The great art of theoretical physics is the revelation of surprising things about reality. Historically there have been many approaches to that art, which have succeeded in different ways. In the early days of physics, people like Galileo and Newton were close to the data and stressed that they were trying to put observed behavior into mathematical terms. They developed some powerful abstract concepts, but by today's standards those concepts were down-to-earth; they were always in terms of things you could touch and feel, or at least see through telescopes. That approach dominated physics, at least through the 19th century. Maxwell's great synthesis of electricity and magnetism and optics, leading to the understanding that light was a form of electricity and magnetism, predicting new kinds of light that we call radio and microwaves, and so forth—that came from a very systematic review of all that was known about electricity and magnetism experimentally, and trying to put it into equations, noticing an inconsistency and fixing it up. That's the kind of classic approach.

In the 20th century, some of the most successful enterprises have looked rather different. Without going into the details it's hard to do justice to all the subtleties, but it's clear that theories like special relativity—especially general relativity—were based on much larger conceptual leaps. In constructing special relativity, Einstein abstracted just two very broad regularities about the physical world: that is, that the laws of physics should look the same if you're moving at a constant velocity and that the speed of light should be a universal constant. This wasn't based on a broad survey of a lot of detailed experimental facts and putting them together; it was selecting a few very key facts and exploiting them conceptually for all they're worth. General relativity even more so: It was trying to make the theory of gravity consistent with the insights of special relativity. This was a very theoretical enterprise, not driven by any specific experimental facts,[1] but it led to a theory that changed our notions of space and time, and did lead to experimental predictions, and to many surprises.

The Dirac equation is a more complicated case. Dirac was moved by broad theoretical imperatives. He wanted to make the existing equation for quantum mechanical behavior of

---

1 Actually, there was a big "coincidence" that Newtonian gravity left unexplained: the equality of inertial and gravitational mass, which was an important guiding clue.

electrons—that's the Schrödinger equation—consistent with special relativity. To do that, he invented a new equation—the Dirac equation—that seemed very strange and problematic, yet undeniably beautiful, when he first found it. That strange equation turned out to require vastly new interpretations of all the symbols in it, interpretations that weren't anticipated. It led to the prediction of antimatter and the beginnings of quantum field theory. This was another revolution that was, in a sense, conceptually driven. On the other hand, what gave Dirac and others confidence that his equation was on the right track was that it predicted corrections to the behavior of electrons in hydrogen atoms that were very specific and agreed with precision measurements. This support forced them to stick with it and find an interpretation to let it be true. So there was important empirical guidance and encouragement from the start.

Our foundational work on QCD [quantum chromodynamics] falls in the same pattern. We were led to specific equations by theoretical considerations, but the equations seemed problematic. They were full of particles that aren't observed—quarks and especially gluons—and didn't contain any of the particles that *are* observed! We persisted with them nevertheless, because they explained a few precision measurements, and that persistence eventually paid off.

In general, as physics has matured in the 20th century, we've realized more and more the power of mathematical considerations of consistency and symmetry to dictate the form of physical laws. We can do a lot with less experimental input. Nevertheless the ultimate standard must be getting experimental output: illuminating reality. How far can aesthetics take you? Should you let that be your main guide, or should you try to assemble and do justice to a lot of specific facts? Different people have different styles; some people try to use a lot of facts and extrapolate a little bit; other people try not to use any facts at all and construct a theory that's so beautiful that it has to be right, and then fill in the facts later. I try to consider both possibilities and see which one is fruitful. What's been fruitful for me is to take salient experimental facts that are somehow striking, or that seem anomalous—don't really fit into our understanding of physics—and try to improve the equations to include just those facts.

My reading of history is that even the greatest advances in physics, when you pick them apart, were always based on a firm empirical foundation and straightening out some anomalies between the existing theoretical framework and some known facts about the world. Certainly QCD was that way; when we developed asymptotic freedom to explain some behaviors of quarks—that they seem to not interact when they're close together—it seemed inconsistent with quantum field theory, but we were able to push and find very specific quantum field theories in which that behavior was consistent, which essentially solved the problem of the strong interaction and has had many fruitful consequences. Axions also—similar thing—a little anomaly: There's a quantity that happens to be very small in the world, but our theories don't explain why it's small. You can change the theories to make them a little more symmetrical—then we do get zero—but that has other consequences. The existence of these new particles rocks cosmology, and they might be the dark matter—I love that kind of thing.

String theory is sort of the extreme of non-empirical physics. In fact, its historical origins were based on empirical observations but wrong ones. String theory was originally based on trying to explain the nature of the strong interactions, the fact that hadrons come in big families, and the idea was that they could be modeled as different states of strings that are spinning around or vibrating in different ways. That idea was highly developed in the late '60s and early '70s, but we put it out of business with QCD, which is a very different theory that turns out to be the correct theory of the strong interaction.

But the mathematics that was developed around that wrong idea, amazingly, turned out to contain—if you do things just right, and tune it up—to contain a description of general relativity and at the same time obeys quantum mechanics. This had been one of the great conceptual challenges of 20th-century physics: to combine the two very different-seeming kinds of theories—quantum mechanics, our crowning achievement in understanding the microworld, and general relativity, which was abstracted from the behavior of space and time in the macroworld. Those theories are of a very different nature, and when you try to combine them you find it's very difficult to make an entirely consistent union of the two. But these evolved string theories seem to do that.

The problems that arise in making a quantum theory of gravity—unfortunately for theoretical physicists who want to focus on them—really only arise in thought experiments of a very high order: thought experiments involving particles of enormous energies, or the deep interior of black holes, or perhaps the earliest moments of the Big Bang, which we don't understand very well. All remote from any practical, doable experiments. It's hard to check the fundamental hypotheses of this kind of idea. The initial hope, when the so-called first string revolution occurred, in the mid-1980s, was that when you actually solved the equations of string theory, you'd find a more or less unique solution, or maybe a handful of solutions, and it would be clear that one of them described the real world.

From these highly conceptual considerations of what it takes to make a theory of quantum gravity, you would be led "by the way" to things we can access and experiment, and it would describe reality. But as time went on, people found more and more solutions with all kinds of different properties, and that hope—that indirectly by addressing conceptual questions you'd be able to work your way down to description of concrete things about reality—has gotten more and more tenuous. That's where it stands today.

My personal style in fundamental physics continues to be opportunistic: To look at the phenomena as they emerge and think about possibilities to beautify the equations that the equations themselves suggest. As I mentioned earlier, I certainly intend to push harder on ideas that I had a long time ago but that still seem promising and still haven't been exhausted: in supersymmetry and axions and even in additional applications of QCD. I'm also always trying to think of new things. For example, I've been thinking about the new possibilities for phenomena that might be associated with this Higgs particle that probably will be discovered at the LHC. I realized something I'd been well aware of at some low level for a long time, but I think now I've realized its profound implications, which is that the Higgs particle uniquely opens a window into phenomena that no other particle within the standard model would be sensitive to. If you look at the mathematics of the standard model, you discover there are possibilities for hidden sectors—things that would interact very weakly with the kind of particles we've had access to so far but would interact powerfully with the Higgs particles. We'll be opening that window. Very recently I've been trying to see if we can get inflation out of the standard model by having the Higgs particle interact in a slightly nonstandard way with gravity. That seems promising too.

Frank Wilczek

Most of my bright ideas will turn out to be wrong, but that's OK. I have fun, and my ego is secure.

In 1993, the Congress of the United States canceled the SSC project, the Superconducting Super Collider, which was under construction near Waxahachie, Texas. Many years of planning, many careers had been invested in that project, also more than $2 billion had already been put into the construction. All that came out of it was a tunnel from nowhere to nothing. Now it's 2009, and a roughly equivalent machine, the Large Hadron Collider, will be coming into operation at CERN near Geneva. The United States has some part in that. It has invested half a billion dollars out of the $15 billion total. But it's a machine that is in Europe, really built by the Europeans; there's no doubt that they have contributed much more. Of course, the information that comes out will be shared by the entire scientific community. So the end result, in terms of tangible knowledge, is the same. We avoided spending the extra money. Was that a clever thing to do?

I don't think so. Even in the narrowest economic perspective, I think it wasn't a clever thing to do. Most of the work that went into this $15 billion was local, locally subcontracted within Europe. It went directly into the economies involved, and furthermore into dynamic sectors of the economy for high-tech industries involved in superconducting magnets, fancy cryogenic engineering, and civil engineering of great sophistication, and of course computer technology. All that know-how is going to pay off much more than the investment in the long run. But even if it weren't the case that purely economically it was a good thing to do, the United States missed an opportunity for national greatness. A hundred or two hundred years from now, people will largely have forgotten about the various spats we got into, the so-called national greatness of imposing our will on foreigners, and they'll remember the glorious expansion of human knowledge that's going to happen at the LHC and the gigantic effort that went into getting it. As a nation, we don't get many opportunities to show history our national greatness, and I think we really missed one there.

Maybe we can recoup. The time is right for an assault on the process of aging. A lot of the basic biology is in place. We know what has to be done. The aging process itself is really the profound aspect of public health; eliminating major diseases, even big ones like cancer or heart disease, would increase life expectancy by only a few years. We really have to get to the root of the process.

Another project on a grand scale would be to search systematically for life in the galaxy. We have tools in astronomy, we can design tools, to find distant planets that might be Earth-like, study their atmospheres, and see if there is evidence for life. It would be feasible, given a national investment of will and money, to survey the galaxy and see if there are additional Earth-like planets that are supporting life. We should think hard about doing things we'll be proud to be remembered for, and think big.

19
Who Cares About Fireflies?

Steven Strogatz

Jacob Gould Schurman Professor of Applied Mathematics,
Cornell University; author, *The Joy of X*

INTRODUCTION by Alan Alda
Actor, writer, director; host of PBS program *Brains on Trial*;
author, *Things I Overheard While Talking to Myself*

Steve Strogatz has worked all his life studying something that
some people thought didn't exist while others thought was too
obvious to mention. It's found in that subtle region—the haze
on the horizon—that smart people, it seems, have always been
intrigued by. He saw something there, and went and looked
closer. What drew him on was a pattern in nature that showed,
surprisingly, that an enormous number of things sync up spontaneously.

His research covered a wide range of phenomena, from sleep
patterns to heart rhythms to the synchronous pulse of Asian fireflies. And in his 2003 book, *Sync*, he drew all these strands (and
many others) together in a way that has the shock of the new.
Even though we may see the moon every night (perhaps not realizing it's an example of sync) it's hard not to be surprised at the
number of things around us—and in us—that must (or must not)
sync up for things to go right.

I've known Steve about thirteen years. We met when I called
him on the phone, wondering if he'd even take my call. I had read

an article of his in *Scientific American* about coupled oscillators. From his first description of Huygens' discovery that pendulum clocks would sync up if they could sense each other's vibrations, I was fascinated, and I hoped he'd tell me more about it. He was surprisingly generous in the face of my hungry, naïve curiosity and we've been friends ever since.

Steve has that quality, like Richard Feynman's, of not only wanting to make every complex thought clear to the average person but, also like Feynman, of actually knowing how. When we were working on the Broadway play *QED*, by Peter Parnell, in which I played Feynman, it's no surprise that we asked Steve to advise us on the physics in the piece.

Please let me introduce you to Steven Strogatz, professor of applied mathematics at Cornell University: my pal, Steve.

Who Cares About Fireflies?

The story of how I got interested in cycles goes back to an epiphany in high school. I was taking a standard freshman science course, Science I, and the first day we were asked to measure the length of the hall. We were told to get down on our hands and knees, put down rulers, and figure out how long the corridor was. I remember thinking to myself, "If this is what science is, it's pretty pointless," and came away with the feeling that science was boring and dusty.

Fortunately I took to the second experiment a little better. The teacher, Mr. diCurcio, said, "I want you to figure out a rule about this pendulum." He handed each of us a little toy pendulum with a retractable bob. You could make it a little bit longer or shorter in clicks in discrete steps. We were each handed a stopwatch and told to let the pendulum swing ten times, and then click, measure how long it takes for ten swings, and then click again, repeating the measurement after making the pendulum a little bit longer. The point was to see how the length of the pendulum determines how long it takes to make ten swings. The experiment was supposed to teach us about graph paper and how to make a relationship between one variable and another, but as I was dutifully plotting the length of time the pendulum took to swing ten times versus its length it occurred to me, after about the fourth or fifth dot, that a pattern was starting to emerge. These dots were falling on a particular curve I recognized because I'd seen it in my algebra class. It was a parabola, the same shape that water makes coming out of a fountain.

I remember having an enveloping sensation of fear. It was not a happy feeling but an awestruck feeling. It was as if this pendulum *knew algebra*. What was the connection between the parabolas in algebra class and the motion of this pendulum? There it was on the graph paper. It was a moment that struck me, and was my first sense that the phrase "law of nature" meant something. I suddenly knew what people were talking about when they said there could be order in the universe and that, more to the point, you couldn't see it unless you knew math. It was an epiphany that I've never really recovered from.

A later experiment in the same class dealt with the phenomenon called resonance. The experiment was to fill a long tube with water up to a certain height, and the rest would contain air. Then you struck a tuning fork above the open end. The tuning fork vibrates at a known frequency—440 cycles per second, the A above Middle C—and the experiment was to raise or lower the water column until it reached the point where a tremendous booming sound would come out. The small sound of the tuning fork would be greatly amplified when the water column was just the right height, indicating that you had achieved resonance. The theory was that the conditions for resonance occur when you have a quarter-wavelength of a sound wave in the open end of the tube, and the point was that by knowing the frequency of the sound wave and measuring the length of the air, you could, sitting there in your high school, derive the speed of sound.

I remember at the time not really understanding the experiment so well, but Mr. diCurcio scolded me and said, "Steve, this is an important experiment, because this is not just about the speed of sound. You have to realize that resonance is what holds atoms together." Again, that gave me that chilling feeling, since I thought I was just measuring the speed of sound, or playing with water in a column, but from diCurcio's point of view this humble water column was a window into the structure of matter. Seeing that resonance could apply to something as ineffable as atomic structure—what makes this table in front of me solid—I was just struck with the unity of nature again, and the idea that these principles were so transcendent that they could apply to everything from sound waves to atoms.

The unity of nature shouldn't be exaggerated, since this is certainly not to claim that everything is the same—but there are certain threads that reappear. Resonance is an idea that we can use to understand vibrations of bridges and to think about atomic structure and sound waves, and the same mathematics applies over and over again in different versions.

There's one other story about Mr. diCurcio that I like. At one time I was reading a biography of Einstein, and diCurcio treated me like I was a full-grown scientist—at thirteen or fourteen. I mentioned to him that Einstein was struck as a young high school student by Maxwell's equations, the laws of electricity and magnetism, and that they made a deep impression on him. I said I couldn't wait until I was old enough, or knew enough math, to know what Maxwell's equations were and to understand them. This being a boarding school, we used to have family-style dinner sometimes, and so he and I were sitting around a big table with several other kids, his two daughters, and his wife, and he was serving mashed potatoes. As soon as I said I would love to see Maxwell's equations sometime, he put down the mashed potatoes and said, "Would you like to see them right now?" And I said, "Yeah, fine." He started writing on a napkin these very cryptic symbols, upside-down triangles and E's and B's and crosses and dots and mumbled, in what sounded like speaking in tongues, "The curl of a curl is grad div minus del squared . . . and from this we can get the wave equation . . . and now we see electricity and magnetism, and can explain what light is." It was one of these awesome moments, and I looked at my teacher in a new way. Here was Mr. diCurcio, not just a high school teacher but someone who knew Maxwell's equations off the top of his head. It gave me the sense that there was no limit to what I could learn from this man.

There was also a Mr. Johnson at this same high school, who was my geometry teacher. Mr. Johnson was an MIT graduate and seemed to know a lot about math. One day he happened to mention that there was a certain problem about triangles that he didn't know how to solve, even though it sounded like every other problem we were hearing about in trigonometry or geometry. The question was, If two angle bisectors of a triangle are the same length, does it have to be an isosceles triangle? That is, do those two angles at the bottom have to be the same? He said he didn't know how to solve it and that in twenty years of teaching he had never seen anyone do so, either. I'd never heard a teacher say there was a geometry problem he didn't know how to do, so I became interested in it. I would be in gym class and someone would throw the ball at me and I wouldn't be paying attention and would drop the ball, still thinking about the angle bisectors. This problem began to obsess me for several months, and I had things that were close to a solution but I could never get it. It was around that time that I learned what research was. I was doing research for the fun of it; there was no grade attached to this, and I didn't even tell anyone I was thinking about it—I just wanted to solve it. One day I thought I had. It was a Sunday morning, and I called up Mr. Johnson and asked him if I could come show him the proof. He said, "Yes, come to my house. Here's where I live." I walked down to his house, and he was still in his pajamas with his kids, and line by line he checked through this proof and said, "You've got it. That's it!" He wasn't really smiling but seemed pleased. He later wrote some special remark to the headmaster of the school that I had done this problem.

I went from Loomis-Chaffee to Princeton, where the path was a little bit bumpy. I started as a freshman taking linear algebra—this is the whiz-kid math course for kids who had done well in high school. The first day, a professor named John Mather walked in, and we couldn't tell if he was a professor or a grad student. He was so shy, had a big long red beard, and slithered along the wall. He didn't really stride into the room—he was practically invisible. Then he began, "The definition of a field, F, is . . ." That was all. He didn't say his name, not "Welcome to Princeton"—nothing. He just began with the definition at the beginning of linear algebra. It was a dreadful experience for me. It was the first time in my life that I understood why people are terrified of math. He came really close to discouraging me from ever wanting to be a mathematician.

Steven Strogatz

The only reason I went on to eventually become one was that my second course was with a great teacher, Eli Stein, who's still at Princeton. It was a course in complex variables, which was a lot like calculus. I always liked calculus in high school, and I suddenly felt like I could do math again, whereas that earlier course on linear algebra was a filter; it had a very fine mesh, and only certain students could get through the holes. What was supposedly being tested was your tendency to think abstractly. Could you come up with the sorts of rigorous proofs that a pure mathematician needs? That's the bread and butter of pure math. The truth probably is that I didn't really have that in me; that wasn't my natural strength. What I really like is math applied to nature—the math of the real world. At the time, I didn't know there was a thing called applied math—I thought it was physics. Now this is what I do. I ended up majoring in math because of the good experience in my sophomore year with Stein.

I'd always been encouraged to be a doctor, but always resisted, because I knew that I wanted to teach math. But in my junior year my parents encouraged me to take some pre-med courses, like biology and chemistry, and even though it was getting to be pretty late to become a doctor I agreed. My brother the lawyer convinced me with a persuasive argument that it was irrational to keep resisting: I wasn't committing to being a doctor, and it wouldn't hurt to learn biology and chemistry. I accepted that, and it made for a pretty hellish year, because I was taking freshman biology, with its lab, freshman chemistry, with another, and organic chemistry, which supposedly depended on freshman chemistry. This is a lot for someone who's not good in the lab.

Even though this pre-med year was a lot of work to be doing alongside a math major, I actually liked the biology and chemistry courses—especially the idea that DNA was a double helix and that this shape would immediately indicate its function. It explained how replication would work. I was perfectly content and even took a pre-med Stanley Kaplan course to prepare for the MCATs. Still, when I got home for spring vacation, my mother got a look at my face and said, "There's something wrong. Something's really bothering you. What's wrong? How do you like school?" I said, "I like it, it's fine, I'm learning good stuff." But she said, "You don't look right. You don't look happy. Something's wrong, what's wrong with you?" I didn't really know. I said, "Maybe I'm tired. I'm working a lot." But she said, "No, something else is really wrong. What are you going to take next year? You'll be a senior." I said, "That's bothering me, because, being a pre-med so late, I'm going to have to take vertebrate physiology, some biochemistry, and all these pre-med courses. Plus I have a senior thesis in the math department, which means that my schedule is going to be so full that I'm not going to be able to take quantum mechanics." And she said, "Why does that matter to you?" I said, "I've been reading about Einstein since I was twelve years old. I love Heisenberg, Niels Bohr, Schrödinger—I could finally understand what they're really talking about. There are no more verbal analogies and metaphors; I can understand what Schrödinger did. I've worked my whole life to get to this point, and I'm ready to know what the Heisenberg uncertainty principle really says, and I'm going to be in medical school, cutting cadavers, and I'm never going to learn it."

So she said, "What if you could just say right now, 'I want to do math. I want to do physics. I want to take quantum mechanics. I'm not going to be a doctor. I want to be the best math teacher and researcher I can be.'" And I just started crying. It was like a tremendous weight had been lifted. We were both laughing

Steven Strogatz

and crying. It was a moment of truth, and I never looked back. I'm very thankful I had such good parents and that I was able to find a passion by denying it for a while. Some people go their whole life and never really figure out what they want to do.

When it was time to do a senior thesis—at Princeton, everyone is required to do one—I wanted to pick something about geometry in nature, although I didn't know exactly what. My adviser, Fred Almgren, who was famous for studying the geometry of soap bubbles, suggested a problem about DNA geometry. Could we understand, for instance, what it might be about DNA that allows it to unwind itself without getting tangled? Given that it's so long, you'd think maybe it would get tangled occasionally and that this would be deadly if it happened in the cell. What keeps it from doing that? By working on this project, I got to learn about DNA and wrote a senior thesis about its geometry. I collaborated with a biochemist to propose a new structure of something called chromatin, which is a mixture of DNA and proteins that makes up our chromosomes. It's the next level of structure after the double helix. We know that the double helix gets wound around little spools of protein called nucleosomes, but no one knows how the nucleosomes themselves become arranged like beads on a string or how that structure is wound up to make chromosomes. My biochemical adviser and I ended up publishing a paper in the *Proceedings of the National Academy of Sciences*, and it was very exhilarating. I was now doing research in mathematical biology, so even though I wasn't a doctor, I was using math in biology about real stuff, about chromosomes.

That was when I realized that I wanted to be an applied mathematician doing mathematical biology. I went to England and studied at Cambridge on a Marshall scholarship and was completely bored with the traditional program—the stuff that G. H. Hardy describes in his book, *A Mathematician's Apology*, known as the Mathematical Tripos. Cambridge has had the same kind of course since Newton, and I was bored with it. One day I found myself wandering into a bookstore across the street, where I picked up a book with the unlikely title, *The Geometry of Biological Time*. I say "unlikely" because I had subtitled my own senior thesis "An Essay in Geometric Biology." I thought I had made up this phrase—geometric biology, the juxtaposition of geometry and biology, shape and life—and here was somebody using practically the same title. So who was this guy, Arthur T. Winfree?

I started reading the book and thought he was nuts; he looked like a crackpot. I didn't know if I should believe that he was serious, because the titles of his chapters have puns in them; he has data from his own mother about her menstrual cycle over many years and it's all about the cycles of living things. At the time, Winfree, who just died on November 5, 2002, was not someone who was really on the map. He was a professor of biology at Purdue. I glanced at the book, put it back on the shelf, and found myself coming back a few days later and read some more, and eventually bought it. Partly because I was so lonesome in England, suffering from culture shock, and partly because this book was so entrancing to me, I started reading and underlining it every day, and fell in love with the vision of living things with many cyclic processes, from cell division to heartbeat to rhythms in the brain, jet lag and sleep rhythms, all described by a single mathematics. This is what Winfree was proposing in his book, and it really is what got me started in the direction of studying synchrony.

We see fantastic examples of synchrony in the natural world all around us. To give a few examples, there were persistent re-

Steven Strogatz

ports when the first Western travelers went to southeast Asia, back to the time of Sir Francis Drake in the 1500s, of spectacular scenes along riverbanks, where thousands upon thousands of fireflies in the trees would all light up and go off simultaneously. These kinds of reports kept coming back to the West and were published in scientific journals, and people who hadn't seen it couldn't believe it. Scientists said this is a case of human misperception, that we're seeing patterns that don't exist, or that it's an optical illusion. How could the fireflies, which are not very intelligent creatures, manage to coordinate their flashings in such a spectacular and vast way?

One theory was that there might be a leader. It sounds ridiculous, because why would there be one special firefly? We don't believe that there's a leader any more, or that there might be atmospheric conditions that cause synchrony, like if a lightning bolt startled every one of them and made them start flashing at the same time, which would cause them to stay together. Synchrony occurs on nights that are perfectly clear. It was only in the 1960s that a biologist named John Buck, from the National Institutes of Health, and his colleagues started to figure out what was really going on, which is that the fireflies are self-organizing. They manage to get in step every night of the year and flash in perfect time for hours upon end with no leader or prompting from the environment, through what is essentially a very mysterious process of emergence. This is a phrase that we hear all the time, but this is emergence in the natural world.

The thinking now is that individual fireflies are able not only to emit flashes but also to respond to the flashes of others—they adjust their own timers. To demonstrate this, Buck and his wife, Elisabeth, who was with him on this trip to Thailand in the mid-'60s, collected bags full of male fireflies, brought them back to their hotel room in Bangkok, and released them in the darkened room. They found that the fireflies were flitting around and crawling on the ceiling and walls and that gradually little pockets of two and then three and four fireflies would be flashing in sync. Later lab experiments showed that by flashing an artificial light at a firefly, you could speed up its rhythm, you could slow it down, make it flash a little later than it would otherwise. The point is that each firefly sends a signal that causes another firefly to speed up or slow down in just the right way so that they end up inevitably coming into sync, no matter how they started.

You might ask why that matters—who cares about fireflies?—but there are lots of reasons why you should care: The first may be that all kinds of applications in technology and medicine depend on this same kind of spontaneous synchronization. In your own heart, you have ten thousand pacemaker cells that trigger the rest of your heart to beat properly, and those ten thousand cells are like the thousands of fireflies. Each one could have its own rhythm. In this case, it's an electrical rhythmic discharge; instead of communicating with light, they're sending electrical currents back and forth to each other—but at an abstract level they're the same. They're individual oscillators that want to have a periodic repetition of their state and can influence each other. They do so in such a way that they conspire to come into sync.

Sometimes, of course, sync is undesirable. If it occurs in your brain on a vast scale it results in epilepsy.

There are other medical and technological applications. The laser, one of the most practical gadgets of our time, depends on light waves in sync, atoms pulsing in unison, all emitting light of the same color and moving in phase, with all the troughs and crests of the light waves perfectly lined up. The light in a laser is

Steven Strogatz

no different from the light coming out of these bulbs overhead, in that the atoms are not really that different; it's the coordination of the atoms that's different. The choreography is the difference, not the dancers.

We see sync all around us. The global positioning system that's been used for satellite-guided weaponry in Iraq consists of twenty-four satellites, each of which has an on-board atomic clock, very well synchronized to a master super clock at NIST, the National Institute of Standards and Technology, in Boulder, Colorado. It's only because of this perfect synchronization, down to a billionth of a second, that we can hit a license plate with a missile. It's a grim example, but then at the same time we see life itself depending on synchrony. Sperm cells beat their tails in unison as they swim toward an egg, communicating through pressure fluctuations in the fluid. I could go on and on. What's so breathtaking about the phenomenon of synchrony is that it occurs at every scale of nature, from the subatomic to the cosmic. It's what may have killed the dinosaurs. I could explain why it was gravitational synchrony that caused certain asteroids to be flung out of the asteroid belt and ultimately to strike our planet, probably extinguishing the dinosaurs and many other creatures. It's one of the most pervasive phenomena in nature, but at the same time one of the most mysterious from a theoretical perspective.

We're used to thinking about entropy, the tendency of complex systems to get more and more disordered, as the dominant force. People always ask me, "Doesn't synchrony violate that? Isn't it against the laws of nature that systems can become spontaneously more ordered?" Of course you can't violate the law of entropy, but there is no contradiction. The point is that the law of entropy applies to a certain class of so-called isolated, or closed systems, where there's no influx of energy from the environment. But that's not what we're talking about when we discuss living things or the Earth. Where systems are far from thermodynamic equilibrium, all bets are off, and we see astonishing feats of self-organization, synchrony just being the simplest such example.

The same laws that give rise to entropy and the tendency of systems to become more and more disordered are the same laws that will account for synchrony. It's just that we don't have a clear enough understanding of the thermodynamics of systems very far from equilibrium to see the connection—but we're getting there. We're learning a lot about spontaneously synchronizing systems, at least in physics. We know how a laser works, and there's nothing that violates entropy about that. For living examples, like heart cells, we have a rough idea about how electrical currents are passed back and forth. But synchrony also touches on some of the deepest mysteries of our time, like consciousness. There's some thought, at least according to some neuroscientists, that what distinguishes consciousness from other forms of brain activity is the synchronized firing of the cells involved at specific frequencies close to 40 cycles a second.

What I like about this whole area of thinking about spontaneous order, or self-organization, is that I really do believe that many of the major unsolved problems of science today have this same character. Architecturally they involve millions of players, whether they're neurons, heart cells, or players in an economy. They're all interacting, influencing each other through complex networks and via complicated interactions, and out of this you sometimes see amazingly organized states. Stuart Kauffman calls this "order for free." Many of us are trying to climb that same mountain from different perspectives. Kauffman and others see it

Steven Strogatz

as a matter of understanding evolution better. When he or Gell-Mann speak of complex adaptive systems the emphasis seems to be on the word "adaptive," suggesting that the key to understanding the mystery is to learn more about natural selection. However, you didn't hear me really talk about evolution at all. I feel like they're barking up a tree that's deeper into the forest. There's a right place to start, and the simplest place to start is to think about problems where evolution plays no part.

I want to think about purely physical systems that are complex in their own right and how, just from the laws of physics, we get these self-organized patterns. It feels to me like I would want to understand that first, before I add the further complication of evolution. We know that's important, but that's starting at the wrong place.

Recently, I keep finding myself wanting to learn more about cancer and what it is about the network of cells or the network of chemical reactions that goes awry in a cancerous cell. There are certainly some cases where a single gene may be screwed up, but I don't believe that all cancers will be explained that way. It's been thirty-five years since Nixon declared war on cancer and we haven't really understood it. Understanding oncogenes is a great start, but that can't be it. Again, it's about choreographies of proteins and genes and the missteps, not just of single dancers but of the way they're moving together. Cancer is somehow a dynamical disease that we won't understand through pure biological reductionist thinking. It's going to take a combination of reductionism to give us the data, and new complex systems theory, supercomputers, and math. I would like to be part of that.

Biologists often emphasize the part that computers will play, and it's true that computers will be indispensable, but there's a third leg, which is good theoretical ideas. It won't be enough to have big computers and great data. You need ideas, and I think those ideas will be expressed in the language of complex-systems mathematics. Although that phrase, "complex systems," has been talked about a lot, I hope people out there appreciate what a feeble state it's in, theoretically speaking. We really don't understand much about it. We have a lot of computer simulations that show stunning phenomena, but where's the understanding? Where's the insight? There are very few cases that we understand, and so that brings me back to synchrony. I like that example of synchrony as a case of spontaneous order because that's one of the few cases we can understand mathematically. If we want to solve these problems, we've got to do the math problems we can do, and we need the simplest phenomena first, and synchrony is among them. It's going to be a long slog to really understand these problems.

Another thought, though, is that we may not need understanding. It could be that understanding is overrated. Perhaps insight is something that's been good for three or four hundred years, since Isaac Newton, but it is not the ultimate end. The ultimate end is really just control of these diseases and avoiding horrible ecological scenarios. If we could get there, even without knowing what we're doing, that would maybe be good enough. Computers might understand it, but we don't have to. There could be a real story here about the overrating of understanding.

In broad strokes, there were hundreds of years after Aristotle when we didn't really understand a whole lot. Once Kepler, Copernicus, and Newton began explaining what they saw through math, there was a great era of understanding, through certain classes of math problems that could be solved. All the mathematics that let us understand laws of physics—Maxwell's equations, thermodynamics, on through quantum theory—all involve a certain class of math problems that we know how to solve completely and thoroughly: that is, linear problems. It's only in the past few decades that we've been banging our heads on the nonlinear ones. Of those, we understand just the smallest ones, using only three or four variables—that's chaos theory. As soon as you have hundreds, or millions, or billions of variables—like in the brain—we don't understand those problems at all. That's what complex systems is supposed to be about, but we're not even close to understanding them. We can simulate them in a computer, but that's not really that different from just watching. We still don't understand.

20
Constructor Theory

David Deutsch

Physicist, University of Oxford; author, *The Beginning of Infinity*; recipient, Edge of Computation Science Prize

Some considerable time ago, we were discussing my idea, new at the time, for constructor theory, which was and is an idea I had for generalizing the quantum theory of computation to cover not just computation but all physical processes. I guessed, and still guess, that this is going to provide a new mode of description of physical systems and laws of physics. It will also have new laws of its own, which will be deeper than the deepest existing theories such as quantum theory and relativity. At the time, I was very enthusiastic about this, and what intervened between then and now is that writing a book took much longer than I expected. But now I'm back to it, and we're working on constructor theory, and, if anything, I would say it's fulfilling its promise more than I expected and sooner than I expected.

One of the first rather unexpected yields of this theory has been a new foundation for information theory. There's a notorious problem with defining information within physics—namely, that on the one hand information is purely abstract, and the original theory of computation as developed by Alan Turing and others regarded computers and the information they manipulate purely abstractly, as mathematical objects. Many mathematicians to this day don't realize that information is physical and that there is no such thing as an abstract computer. Only a physical object can compute things.

On the other hand, physicists have always known that in order to do the work that the theory of information does within physics—such as informing the theory of statistical mechanics, and thereby thermodynamics, the second law of thermodynamics—information has to be a physical quantity. And yet information is independent of the physical object it resides in.

I'm speaking to you now: Information starts as some kind of electrochemical signals in my brain, and then it gets converted into other signals in my nerves and then into sound waves and then into the vibrations of a microphone, mechanical vibrations, then into electricity, and so on, and presumably will eventually go on the Internet. This "something" has been instantiated in radically different physical objects that obey different laws of physics. Yet in order to describe this process, you have to refer to the thing that has remained unchanged throughout the process, which is only the information rather than any obviously physical thing like energy or momentum.

The way to get this substrate independence of information is to refer it to a level of physics that is below and more fundamental than things like laws of motion, that we have been used thinking of as near the lowest, most fundamental level of physics. Constructor theory is that deeper level of physics, physical laws, and physical systems, more fundamental than the existing prevailing conception of what physics is—namely, particles and waves and space and time and an initial state and laws of motion that describe the evolution of that initial state. What led to this hope for this new kind of foundation for the laws of physics was really the quantum theory of computation.

I had thought for a while that the quantum theory of computation is the whole of physics. The reason why it seemed reasonable to think that was that a universal quantum computer can simulate any other finite physical object with arbitrary accuracy, and that means that the set of all possible motions, which is computations, of a universal computer corresponds to the set of all possible motions of anything. There's a certain sense in which studying the universal quantum computer is the same thing as studying every other physical object. It contains all possible motions of all possible physical objects within its own possible diversity. I used to say that the quantum theory of computation is the whole of physics because of this property. But then I realized that that isn't quite true and there's an important gap in that connection. Namely, although the quantum computer can simulate any other object and can represent any other object so that you can study any object via its characteristic programs, what the quantum theory of computation *can't* tell you is, which program corresponds to which physical object.

This might sound like an inessential technicality, but it's actually of fundamental importance, because not knowing which abstraction in the computer corresponds to which object is a little bit like having a bank account and the bank telling you, "Oh, your balance is some number." Unless you know what number it is, you haven't really expressed the whole of the physical situation of you and your bank account. Similarly, if you're told only that your physical system corresponds to some program of the quantum computer and you haven't said which, then you haven't specified to the whole of physics.

Then I thought, "What we need is a generalization of the quantum theory of computation that does say that—that assigns to each program the corresponding real object." That was an early conception of constructor theory—making it directly a generalization of the theory of computation. But then I realized that that's not quite the way to go, because that still tries to cast

David Deutsch

constructor theory within the same mold as all existing theories, and therefore it wouldn't solve this problem of providing an underlying framework. It still would mean that, just as a program has an initial state and then laws of motion—that is, the laws of the operation of the computer—and then a final state which is the output of the computer, so that way of looking at constructor theory would have simply been a translation of existing physics. It wouldn't have provided anything new.

The new thing, which I think is the key to the fact that constructor theory delivers new content, was that the laws of constructor theory are *not* about an initial state, laws of motion, final state, or anything like that. They're just about which transformations are possible and which are impossible. The laws of motion, and that kind of thing, are indirect remote consequences of just saying what's possible and what's impossible. Also, the laws of constructor theory are not about the constructor; they're not about how you do it, only whether you *can* do it, and this is analogous to the theory of computation. The theory of computation isn't about transistors and wires and input/output devices and so on. It's about which transformations of information are possible and which aren't possible. Since we have the universal computer, we know that each possible one corresponds to a program for a universal computer, but the universal computer can be made in lots of different ways. How you make it is inessential to the deep laws of computation.

In the case of constructor theory, what's important is which transformations of physical objects are possible and which are impossible. When they're possible, you'll be able to do them in lots of different ways, usually. When they're impossible, that will always be because some law of physics forbids them, and that's why, as Karl Popper said, the content of a physical theory, of any scientific theory, is in what it forbids and also in how it explains what it forbids.

If you have this theory of what is possible and what is impossible, it implicitly tells you what all the laws of physics are. That very simple basis is proving very fruitful already, and I have great hopes that various niggling problems and notorious difficulties in existing formulations of physics will be solved by this single idea. It may well take a lot of work to see how, but that's what I expect, and I think that's what we're beginning to see. This is often misunderstood as claiming that only the scientific theories are worth having. Now that, as Popper once remarked, is a silly interpretation. For example, Popper's own theory is a philosophical theory. He certainly wasn't saying that that was an illegitimate theory.

In some ways, this theory, just like quantum theory and relativity and anything that's fundamental in physics, overlaps with philosophy. So having the right philosophy—which is the philosophy of Karl Popper, basically—though not essential, is extremely helpful to avoid going down blind alleys. Popper, I suppose, is most famous for his criterion of demarcation between science and metaphysics: Scientific theories are those that are in principle testable by experiment, and what he called metaphysical theories—I think they would be better called philosophical theories—are the ones that can't.

Being testable is not as simple a concept as it sounds. Popper investigated in great detail and laid down principles that lead me to the question, "In what sense is constructor theory testable?" Constructor theory consists of a language in which to express other scientific theories—well, that can't be true or false, it can only be convenient or inconvenient—but also laws. But these laws are not about physical objects, they're laws about

David Deutsch

other laws. They say that other laws have to obey constructor theoretic principles.

That raises the issue of how you can test a law about laws, because if it says that laws have to have such-and-such a property, you can't actually go around and find a law that doesn't have that property, because experiment could never tell you that that law was true. Fortunately, this problem has been solved by Popper. You have to go indirectly, in the case of these laws about laws. I want to introduce the terminology that laws about laws should be called principles. A lot of people already use that kind of terminology, but I'd rather make it standardized.

For example, take the principle of the conservation of energy, which is a statement that all laws have to respect the conservation of energy. Perhaps it's not obvious to you, but there is no experiment that would show a violation of the conservation of energy, because if somebody presented you with an object that produced more energy than it took in, you could always say, "Ah, well, that's due to an invisible thing, or a law of motion that's different from what we think, or maybe the formula for energy is different for this object than what we thought," so there's no experiment that could ever refute it.

And in fact in the history of physics the discovery of the neutrino was made by exactly that method. It appeared that the law of conservation of energy was not being obeyed in beta decay, and then Pauli suggested that maybe the energy was being carried off by an invisible particle that you couldn't detect. It turned out that he was right, but the way you have to test that is not by doing an experiment on beta decay but by seeing whether the theory, the law that says that the neutrino exists, is successful and independently testable. It's the testability of the law that the principle tells you about, that in effect provides the testability of the principle.

One thing I think is important to stress about constructor theory is when I say we want to reformulate physics in terms of what can and can't be done, that sounds like a retreat into operationalism, or into positivism, or something: that we shouldn't worry about the constructor—that is, the thing that does the transformation—but only ~~in~~ the input and output and whether they're compatible. But actually that's not how it works in constructor theory.

Constructor theory is all about how it comes about; it just expresses this in a different language. I'm not very familiar with the popular idea of cybernetics that came about a few decades ago, but I wouldn't be surprised if those ideas, which proved at the time not to lead anywhere, were actually an early avatar of constructor theory. If so, we'll be able to see that only in hindsight, because some of the ideas of constructor theory are really impossible to have until you have a conceptual framework that's post–quantum theory of computation—i.e., after the theory of computation has been explicitly incorporated into physics, not just philosophically. That's what the quantum theory of computation did.

I'm not sure whether von Neumann used the term "constructor theory"—or did he just call it the universal constructor? Von Neumann's work in the 1940s is another place where constructor theory could be thought to have its antecedents. But von Neumann was interested in different issues. He was interested in how living things can possibly exist given what the laws of physics are. This was before the DNA mechanism was discovered.

He was interested in issues of principle—how the existence of a self-replicating object was even consistent with the laws of physics as we know them. He realized that there was an underlying logic and underlying algebra, something that in which

David Deutsch

one could express this and show what was needed. He actually solved the problem of how a living thing could possibly exist, basically by showing that it couldn't possibly work by literally copying itself. It had to have within it a code, a recipe, a specification—or computer program, as we would say today—specifying how to build it, and therefore the self-replication process had to take place in two stages. He did this all before the DNA system was known, but he never got any further, because, from my perspective, he never got anywhere with constructor theory or with realizing that this was all at the foundations of physics rather than just the foundations of biology, because he was stuck in this prevailing conception of physics as being about initial conditions, laws of motion, and final state. Where, among other things, you have to include the constructor in your description of a system, which means that you don't see the laws about the transformation for what they are.

When he found he couldn't make a mathematical model of a living object by writing down its equations on a piece of paper, he resorted to simplifying the laws of physics and then simplifying them again and again, and eventually he invented the whole field we now call cellular automata. It's a very interesting field, but it takes us away from real physics, because it abstracts away the laws of physics. What I want to do is go in the other direction—to integrate it with laws of physics, not as they are now but with the laws of physics that have an underlying algebra that resembles, or is a generalization of, the theory of computation.

Several strands led toward this. I was lucky enough to be placed in more than one of them. The main thing was that starting with Turing, and then Rolf Landauer, who was a lone voice in the 1960s saying that computation is physics . . . Because the theory of computation to this day is regarded by mathematicians as being about abstractions rather than as being about physics. Landauer realized that the concept of a purely abstract computer doesn't make sense, and the theory of computation has to be a theory of what physical objects can do to information. Landauer focused on what restrictions the laws of physics imposed on what kinds of computation can be done. Unfortunately, that was the wrong way around, because, as we later discovered, the most important thing about the relationship of physics with computation, and the most striking thing, is that quantum theory—i.e., the deepest laws of physics that we know—permits new modes of computation that wouldn't be possible in classical physics. Once you have established the quantum theory of computation, you've got a theory of computation that is wholly within physics, and it's then natural to try to generalize that, which is what I wanted to do. So that's one of the directions.

Von Neumann was motivated, really, by theoretical biology rather than theoretical physics. Another thing that I think inhibited von Neumann from realizing that his theory was fundamental physics was that he had the wrong idea about quantum theory. He had settled for, and was one of the pioneers of, building a cop-out version of quantum theory that made it into just an operational theory, where you would use quantum theory just to work out and predict the outcomes of experiments rather than express the laws of how the outcome comes about. That was one of the reasons von Neumann never thought of his own theory as being a generalization of quantum theory—because he didn't really take quantum theory seriously. His contribution to quantum theory was to provide this set of von Neumann rules that allows you to use the theory in practice without ever wondering what it means.

I came from a different tradition of thinking, via Hugh Ever-

David Deutsch

ett and Karl Popper, in their different ways. Both of them insisted that scientific theories are about what is really there and why observations come about, not just predicting what the observations are. Therefore, I couldn't be satisfied with just an operational version of quantum mechanics.

I had to embrace the Everett, or many-universes, interpretation of quantum mechanics from the present point of view. The key thing about that is that it's a realistic theory, as the philosophers say. That is, it's a theory that purports to describe what really happens rather than just our experiences of what happens. Once you think of quantum theory that way, it's only a very small step to realizing, first of all, that computation is the quantum theory of computation, which was my earlier work, and then that the quantum theory of computation is not sufficient to provide the foundation for the whole of physics. So, what's the rest? Well, the rest is constructor theory.

What's needed in constructor theory is to express it in terms that can be integrated with the rest of physics, formulas, equations, because only then can it make contact with other scientific theories. The principles of constructor theory then constrain the laws of other theories, which I call subsidiary theories now. The constructor theory is the deepest theory, and everything else is subsidiary to it. It constrains them, and that, then, leads to contact with experiment.

The key thing to do, apart from guessing what the actual laws are, is to find a way of expressing them. The first item on the agenda, then, is to set up a constructor-theoretic algebra, which is an algebra in which you can do two things. One is to express any other scientific theory in terms of what transformations can or cannot be performed. The analog in the prevailing formulation of physics would be something like differential equations, but in constructor theory it will be an algebra. And then to use that algebra also to express the laws of constructor theory, which won't be expressed in terms of subsidiary theories. They will just make assertions about subsidiary theories.

Chiara Marletto, who's a student I'm working with, and I are working on that algebra. It's a conceptual jolt to think in terms of it rather than in the terms that have been traditional in physics for the last few decades. We try and think what it means, find contradictions between different strands of thought about what it means, realize that the algebra and the expressions that we write in the algebra don't quite make sense, change the algebra, see what that means and so on. It's a process of doing math, doing algebra, by working out things, interleaved with trying to understand what those things mean.

This rather mirrors how the pioneers of quantum theory developed their theory, too. In fact one of the formulations of quantum theory—namely, matrix mechanics, as invented by Heisenberg and others—isn't based on the differential-equation paradigm but is more algebraic, and it, in fact, is another thing that can be seen as a precursor of constructor theory. We haven't yet succeeded in making a viable algebra, but even with the rudimentary form of it we have now, we've got some remarkable results—this was mainly done by Chiara—which have almost magically provided a deeper foundation for information theory than was possible within physics before.

Like all fundamental theories, it's difficult to predict what effect they will have, precisely because they're going to change things at a fundamental level. But there's one big thing I'm pretty sure the constructor-theoretic way of looking at physics has to offer our worldview in terms of everyday life—and that is optimism. "Optimism," in my terminology, doesn't mean expecting that things will turn out well all the time. It's this very specific thing that I think captures the historical, philosophical trend of what optimism has meant if you remove the nonsense. Namely, the optimistic view is not that problems will not occur but that all problems and all evils are caused by lack of knowledge. And the converse of that is that all evils are soluble, given the right knowledge. And knowledge can be found by the methods of conjecture, criticism, and so on, which we know is the way of creating knowledge.

Although this sounds like a statement at a very human level, because it's about knowledge and evils and being able to do things, and so on, it's directly linked with constructor theory at the most fundamental level, because of the fundamental dichotomy in constructor theory, which claims that the whole of science is to be formulated in terms of the difference between transformations that are possible and those that are impossible and there isn't a third possibility. That's the thing. The whole thing doesn't work if there's a third possibility.

If a task, a transformation, is impossible, then constructor theory says it must be because there's some law of physics that makes it impossible. Conversely, if there isn't a law of physics that makes it impossible, then it's possible. There's no third possibility. What does "possible" mean? In the overwhelming majority of cases, though some things are possible because they happen spontaneously, things that are possible are possible because the right knowledge embodied in the right physical object would make them happen. Since the dichotomy is between that which is forbidden by the laws of physics and that which is possible with the right knowledge, and there isn't any other possibility, this tells us that all evils are due to lack of knowledge.

This is counterintuitive. It's contrary to conventional wisdom, and it's contrary to our intuitive, or at least culturally intuitive, way of looking at the world. I find myself grasping for a third possibility. Isn't there something that we can't do even though there's no actual law of physics that says we won't be able to do it? Well, no, there can't be. This is built into constructor theory. There's no way of getting around it, and I think once you've seen that it's at the foundations of physics, it becomes more and more natural. It becomes more and more sort of obvious, in the sense of "It's weird but what else could it be?"

It's rather like the intuitive shift that comes from realizing that people in Australia really are upside-down compared with us, and they really are down there through the Earth. One can know this intellectually, but to actually think in those terms takes an effort. It's something that we all learn at some point to accept intellectually if we're rational. But then to incorporate that into our worldview changes us. It changes us, for instance, because whole swaths of supernatural thinking are made impossible by truly realizing that the people in Australia are upside-down. And, similarly, whole swaths of irrational thinking are made impossible by realizing that, in the sense I've just described, there is no third possibility between being able to do it if we have the right knowledge and its being forbidden by the laws of physics.

The stereotype of how new ideas get into fundamental science is: First, somebody has the idea. Everyone thinks he's crazy, and eventually he's vindicated. I don't think it happens like that very often. There are cases where it does, but much more often—and this is my own experience, when I've had new ideas—it's not that people say, "You're crazy; that can't be true." They say, "Yes, that's nice, well done," and then they go off and ignore it. And then, eventually, people say, "Oh, well, maybe it leads to this," or "maybe it leads to that," and "Maybe it's worth working on. Maybe it's fruitful," and then eventually they work more on it.

This has happened in several of the things that I've done, and this is what I would expect to happen with constructor theory. I haven't had anyone tell me that this is crazy and it can't be right, but I'm certainly in the stage of most people, or most physicists, saying, "Well, that's nice. That's interesting. Well done," and then going away and ignoring it.

No one else is actually working on it at the moment. Several of our colleagues have expressed something between curiosity and substantial interest, which may well go up as soon as we have results. At the moment, there's no publication. I've submitted a philosophical paper, which hasn't even been published yet. When that comes out, it'll get a wider readership. People will understand what it's about, but while in philosophy you can write a paper that just has hopes in it, or interpretations of things, in physics you need results. When we have our first few papers that have results, I think that an exponential process of people working on it will begin—if it's true. Of course, it might be that some of these results are of the form "it can't be true," in which case it will end up as an interesting footnote to the history of physics.

I had to write the philosophical paper first because there's quite a lot of philosophical foundation to constructor theory, and to put that into a physics paper would have simply made it too long and, to physicists, too boring. So I had to write something that we can refer to. It's the philosophical paper first, and then the next thing was going to be constructor-theory algebra, which is the language and formalism and showing how both old laws and new constructor-theoretic laws can be expressed. But now it's likely that the first paper on constructor theory will be constructor-theoretic information theory, because it's yielded unexpectedly good results there.

We're talking about the foundations of physics here, so the question is whether the theory is consistent, whether it's fruitful, whether it leads to new discoveries. These foundational theories are of interest to people who like foundational theories, but their usefulness comes in their fruitfulness later.

Quantum theory is, again, a very good example. Almost nobody was actually interested in quantum theory except a few people who work on the foundations of quantum theory. But now several decades after its discovery, everybody who works on microchips, or everyone who works on information or cryptography, and so on, has to use quantum mechanics; and everybody who wants to understand their position in the universe—what we are—has to take a view of what quantum theory tells us about what we are.

For example, you have to take a view about whether it's really true that we exist in vast numbers of parallel copies, some of them slightly different, some of them the same, as I think quantum mechanics inevitably leads to—or not. But there's no rational way of not taking a position on that issue. Although apart from the issue of optimism, which is an unexpectedly direct connection to the everyday level, we can't tell at the moment what constructor theory will tell us about ourselves and our position in the universe, and what every reasonable person should know, until we work out what the theory says—which we can't do until we work it out properly within the context of theoretical physics.

I'm interested, basically, in anything that's fundamental. It's not confined to fundamental physics, but for me that's what it all revolves around. In the case of constructor theory, how this is

David Deutsch

going to develop totally depends on what the theory turns out to say, and even more fundamentally whether it turns out to be true. If it turns out to be false that one cannot build a foundation to physics in the constructor-theoretic way, that will be extremely interesting, because that will mean that whole lines of argument that seemed to make it inevitable that we need a constructor theory are actually wrong, and whole lines of unification that seem to connect different fields don't connect them, and yet therefore they must be connected in some other way, because the truth of the world has to be connected.

If it turns out to be wrong, the chances are it will be found to be wrong long before it's falsified. This, again, is the typical way of scientific theories. What gets the headlines is if you do an experiment and you predict a certain particle and it doesn't appear and then you're proved wrong. But actually the overwhelming majority of scientific theories are proved wrong long before they ever get tested. They're proved wrong by being internally inconsistent or being inconsistent with other theories that we believe to be true, or most often they're proved wrong by not doing the job of explanation they were designed to do. So if you have a theory that's supposed to, for example, explain the second law of thermodynamics and why there is irreversibility when the fundamental laws of physics are reversible, and then you find by analyzing this theory that it doesn't actually do that, then you don't have to bother to test it, because it doesn't address the problem it was designed to address. If constructor theory turns out to be false, I think it's overwhelmingly likely that it will be by that method— that it just doesn't do this unification job, or foundational job, that it was designed to do.

Then we would have to learn the lesson of how it turned out to be wrong. Turning out to be wrong is not a disgrace. It's not like in politics, where if you lose the election then you've lost. In science, if your idea that looked right turns out to be wrong, you've learned something.

One of the central philosophical motivations for why I do fundamental physics is that I'm interested in what the world is like—that is, not just the world of our observations, what we see, but the invisible world, the invisible processes and objects that bring about the visible. Because the visible is only the tiny, superficial, and parochial sheen on top of the real reality, and the amazing thing about the world and our place in it is that we can discover the real reality.

We can discover what's at the center of stars even though we've never been there. We can find out that those cold, tiny objects in the sky that we call stars are actually million-kilometer, white, hot, gaseous spheres. They don't look like that. They look like cold dots, but we know different. We know that the invisible reality is there giving rise to our visible perceptions.

That science has to be about that has been for many decades a minority and unpopular view among philosophers and, to a great extent, regrettably even among scientists. They have taken the view that science, just because it's characterized by experimental tests, has to be only about experimental tests, but that's a trap. If that were so, it would mean that science is only about humans, and not even everything about humans but about human experience only. It's solipsism. It's purporting to have a rigorous, objective worldview that only observations count, but ending up by its own inexorable logic as saying that only human experience is real, which is solipsism.

I think it's important to regard science not as an enterprise for the purpose of making predictions but as an enterprise for the purpose of discovering what the world is really like, what is really

David Deutsch

there, how it behaves and why. Which is tested by observation. But it's absolutely amazing that the tiny little parochial and weak and error-prone access that we have to observations is capable of testing theories and knowledge of the whole of reality, which has tremendous reach far beyond our experience. And yet we know about it. That's the amazing thing about science. That's the aspect of science that I want to pursue.

21
A Theory of Roughness

Benoit Mandelbrot (1924-2010)

Mathematician, Yale University; author, *The Fractal Geometry of Nature,* d. October 14, 2010

INTRODUCTION by John Brockman

During the 1980s, Benoit Mandelbrot accepted my invitation to give a talk before The Reality Club. The evening was the toughest ticket in the ten-year history of live Reality Club events during that decade; it seemed like every artist in New York had heard about it and wanted to attend. It was an exciting, magical evening. I stayed in touch with Mandelbrot and shared an occasional meal with him every few years, always interested in what he had to say.

He didn't make it easy. While his ideas were complicated but understandable, his French accent could make comprehension more difficult. His death in 2010 was a blow to anyone who values individuals who dedicate their lives to a scientific observation of the natural world. While the inclusion of this conversation in a book about the universe might seem tangential to some readers, it is in part a tribute to him and to his tireless capacity for what our efforts to understand the universe require: Reinvention.

Mandelbrot is best known as the founder of fractal geometry, which impacts mathematics, diverse sciences, and arts. Here he continues to push the envelope, with his theory of roughness. "There is a joke that your hammer will always find nails to hit," he says. "I find that perfectly acceptable. The hammer I crafted is the first effective tool for all kinds of roughness and nobody will deny that there is at least some roughness everywhere."

A Theory of Roughness

There is a saying that every nice piece of work needs the right person in the right place at the right time. For much of my life, however, there was no place where the things I wanted to investigate were of interest to anyone. So I spent much of my life as an outsider, moving from field to field, and back again, according to circumstances. Now that I near eighty, write my memoirs, and look back, I realize with wistful pleasure that on many occasions I was ten, twenty, forty, even fifty years ahead of my time. Until a few years ago, the topics in my PhD were unfashionable, but they are very popular today.

My ambition was not to create a new field, but I would have welcomed a permanent group of people having interests close to mine and therefore breaking the disastrous tendency toward increasingly well-defined fields. Unfortunately, I failed on this essential point, very badly. Order doesn't come by itself. In my youth I was a student at Caltech while molecular biology was being created by Max Delbrück, so I saw what it means to create a new field. But my work did not give rise to anything like that. One reason is my personality—I don't seek power and do not run around. A second is circumstances—I was in an industrial laboratory, because academia found me unsuitable. Besides, creating close organized links between activities which otherwise are very separate might have been beyond any single person's ability.

That issue is important to me now, in terms of legacy. Let me elaborate. When I turned seventy, a former postdoc organized a festive meeting in Curaçao. It was superb, because of the participation of mathematician friends, physicist friends, engineering friends, economist friends, and many others. Geographically, Curaçao is out of the way, hence not everybody could make it, but every field was represented. Several such meetings had been organized since 1982. However, my enjoyment of Curaçao was affected by a very strong feeling that this was going to be the last such common meeting. My efforts over the years had been successful to the extent—to take an example—that fractals made many mathematicians learn a lot about physics, biology, and economics. Unfortunately, most were beginning to feel they had learned enough to last for the rest of their lives. They remained mathematicians, had been changed by considering the new problems I raised, but largely went their own way.

Today, various activities united at Curaçao are again quite separate. Notable exceptions persist, to which I shall return in a moment. However, as I was nearing eighty, a Curaçao-like meeting was not considered at all. Instead, the event is being celebrated by more than half a dozen specialized meetings in diverse locations. The most novel and most encouraging one will be limited to very practical applications of fractals, to issues concerning plastics, concrete, the Internet, and the like.

For many years I had been hearing the comment that fractals make beautiful pictures but are pretty useless. I was irritated, because important applications always take some time to be revealed. For fractals, it turned out that we didn't have to wait very long. In pure science, fads come and go. To influence basic big-budget industry takes longer, but hopefully also lasts longer.

To return to and explain how fractals have influenced pure mathematics, let me say that I am about to spend several weeks at the Mittag-Leffler Institute at the Swedish Academy of Sciences. Only twenty-five years ago, I had no reason to set foot there, except to visit the spectacular library. But, as it turned out, my work has inspired three apparently distinct programs at this Institute.

The first was held in the 1980s when the Mandelbrot Set was a topic of a whole year of discussion. It may not be widely appreciated that the discovery of that set had consisted in empowering the eye again, in inspecting pictures beyond counting and on their basis stating a number of observations and conjectures to which I drew the mathematicians' attention. One of my conjectures was solved in six months, a second in five years, a third in ten. But the basic conjecture, despite heroic efforts rewarded by two Fields Medals, remains a conjecture, now called MLC: the Mandelbrot Set is locally connected. The notion that these conjectures might have been reached by pure thought—with no picture—is simply inconceivable.

The next Mittag-Leffler year I inspired came six years ago and focused on my "4/3" conjecture about Brownian motion. Its discovery is characteristic of my research style and my legacy, hence deserves to be retold.

Scientists have known Brownian motion for centuries, and the mathematical model provided by Norbert Wiener is a marvelous pillar at the very center of probability theory. Early on, scientists had made pictures both of Brownian motion in nature and of Wiener's model. But this area developed like many others in mathematics and lost all contact with the real world.

My attitude has been totally different. I always saw a close kinship between the needs of "pure" mathematics and a certain hero of Greek mythology, Antaeus. The son of Earth, he had to touch the ground every so often in order to reestablish contact with his Mother; otherwise his strength waned. To strangle him, Hercules simply held him off the ground. Back to mathematics. Separation from any down-to-earth input could safely be complete for long periods—but not forever. In particular, the mathematical study of Brownian motion deserved a fresh contact with reality.

Seeking such a contact, I had my programmer draw a very big sample motion and proceeded to play with it. I was not trying to implement any preconceived idea, simply actively "fishing" for new things. For a long time, nothing new came up. Then I conceived an idea that was less scientific than aesthetic. I became bothered by the fact that when a Brownian motion has been drawn from time 0 to time 1, its two end portions and its middle portion follow different rules. That is, the whole is not homogeneous and exhibits a certain lack of inner symmetry, a deficit of beauty.

This triggered the philosophical prejudice that when you seek some unspecified and hidden property, you don't want extraneous complexity to interfere. In order to achieve homogeneity, I decided to make the motion end where it had started. The resulting motion biting its own tail created a distinctive new shape I call Brownian cluster. Next, the same purely aesthetic consideration led to further processing. The continuing wish to eliminate extraneous complexity made me combine all the points that cannot be reached from infinity without crossing the Brownian cluster. Painting them in black sufficed, once again, to create something quite new, resembling an island. Instantly, it became apparent that its boundary deserved to be investigated. Just as instantly, my long previous experience with the coastlines of actual islands on Earth came in handy and made me suspect that the boundary of Brownian motion has a fractal dimension equal to 4/3. The fractal dimension is a concept that used to belong to well-hidden mathematical esoteric. But in the previous decades I had tamed it into becoming an intrinsic qualitative measure of roughness.

Empirical measurement yielded 1.3336, and on this basis my

Benoit Mandelbrot (1924–2010)

1982 book, *The Fractal Geometry of Nature*, conjectured that the value of 4/3 is exact. Mathematician friends chided me: Had I told them before publishing, they could have quickly provided a fully rigorous proof of my conjecture. They were wildly over-optimistic, and a proof turned out to be extraordinarily elusive. A colleague provided a numerical approximation that fitted 4/3 to about fifteen decimal places, but an actual proof took eighteen years and the joining of contributions of three very different scientists. It was an enormous sensation in the year 2000. Not only the difficult proof created its own very active subfield of mathematics but it affected other, far removed, subfields by automatically settling many seemingly unrelated conjectures. An article in *Science* magazine reported to my great delight a comment made at a major presentation of the results, that this was the most exciting thing in probability theory in twenty years. Amazing things started happening, and the Mittag-Leffler Institute organized a full year to discuss what to do next.

Today, after the fact, the boundary of Brownian motion might be billed as a "natural" concept. But yesterday this concept had not occurred to anyone. And even if it had been reached by pure thought, how could anyone have proceeded to the dimension 4/3? To bring this topic to life, it was necessary for the Antaeus of mathematics to be compelled to touch his Mother Earth, if only for one fleeting moment.

Within the mathematical community, the MLC and 4/3 conjectures had a profound effect—witnessed recently when the French research council, CNRS, expressed itself as follows. "Mathematics operates in two complementary ways. In the 'visual' one, the meaning of a theorem is perceived instantly on a geometric figure. The 'written' one leans on language, on algebra; it operates in time. Hermann Weyl wrote that 'the angel of geometry and the devil of algebra share the stage, illustrating the difficulties of both.'" I, who took leave from French mathematics at age twenty because of its rage against images, could not have described it better. Great to be alive when these words come from that pen. But don't forget that in the generations between Hermann Weyl (1885-1955) and today—the generations of my middle years—the mood had been totally different.

Back to cluster dimension. At IBM, where I was working at the time, my friends went on from the Brownian to other clusters. They began with the critical percolation cluster, which is a famous mathematical structure of great interest in statistical physics. For it, an intrinsic complication is that the boundary can be defined in two distinct ways, yielding 4/3, again, and 7/4. Both values were first obtained numerically but by now have been proven theoretically, not by isolated arguments serving no other purpose but in a way that has been found very useful elsewhere. As this has continued, an enormous range of geometric shapes, so far discussed physically but not rigorously, became attractive in pure mathematics, and the proofs were found to be very difficult and very interesting.

The third meeting that my work inspired at the Mittag-Leffler Institute of the Swedish Academy will take place this year. Its primary concern will be a topic I have already mentioned, the mathematics of the Internet.

This may or may not have happened to you, but some non-negligible proportion of email gets lost. Multiple identical messages are a pest, but the sender is actually playing it safe, for the good reason that in engineering everything is finite. There is a very complicated way in which messages get together, separate, and are sorted. Although computer memory is no longer expensive, there's always a finite-size buffer somewhere. When a big

Benoit Mandelbrot (1924-2010)

piece of news arrives, everybody sends a message to everybody else, and the buffer fills. If so, what happens to the messages? They're gone, just flow into the river.

At first, the experts thought they could use an old theory that had been developed in the 1920s for telephone networks. But as the Internet expanded, it was found that this model won't work. Next, they tried one of my inventions from the mid-1960s, and it wouldn't work either. Then they tried multifractals, a mathematical construction I had introduced in the late 1960s and into the 1970s. Multifractals are the sort of concept that might have been originated by mathematicians for the pleasure of doing mathematics, but in fact it originated in my study of turbulence and I immediately extended it to finance. To test new Internet equipment, one examines its performance under multifractal variability. This is even a fairly big business, from what I understand. How could it be that the same technique applies to the Internet, the weather, and the stock market? Why, without particularly trying, am I touching so many different aspects of many different things?

A recent, important turn in my life occurred when I realized that something I have long been stating in footnotes should be put on the marquee. I have engaged myself, without realizing it, in undertaking a theory of roughness. Think of color, pitch, heaviness, and hotness. Each is the topic of a branch of physics. Chemistry is filled with acids, sugars, and alcohols; all are concepts derived from sensory perceptions. Roughness is just as important as all those other raw sensations, but was not studied for its own sake.

In 1982, a metallurgist approached me, with the impression that fractal dimension might provide at long last a measure of the roughness of such things as fractures in metals. Experiments confirmed this hunch, and we wrote a paper for *Nature* in 1984. It brought a big following and actually created a field concerned with the measurement of roughness. Recently, I have moved the contents of that paper to page 1 of every description of my life's work.

Those descriptions have repeatedly changed, because I was not particularly precocious, but I'm particularly long-lived and continue to evolve even today. Above a multitude of specialized considerations, I see the bulk of my work as having been directed toward a single overarching goal: to develop a rigorous analysis for roughness. At long last, this theme has given powerful cohesion to my life. Earlier on, since my PhD thesis in 1952, the cohesion had been far more flimsy. It had been based on scaling—that is, on the central role taken by so-called power-law relations.

For better or worse, none of my acquaintances has or had a similar story to tell. Everybody I have known has been constantly conscious of working in a pre-existing field or in one being consciously established. As a notable example, Max Delbrück was first a physicist and then became the founder of molecular biology, a field he always understood as extending the field of biology. To the contrary, my fate has been that what I undertook was fully understood only after the fact, very late in my life.

To appreciate the nature of fractals, recall Galileo's splendid manifesto that "Philosophy is written in the language of mathematics and its characters are triangles, circles and other geometric figures, without which one wanders about in a dark labyrinth." Observe that circles, ellipses, and parabolas are very smooth shapes and that a triangle has a small number of points of irregularity. Galileo was absolutely right to assert that in science those shapes are necessary. But they have turned out not to be sufficient, "merely" because most of the world is of infinitely great roughness and complexity. However, the infinite sea of complexity includes two islands: one of Euclidean simplicity, and also a

Benoit Mandelbrot (1924-2010)

second of relative simplicity in which roughness is present but is the same at all scales.

The standard example is the cauliflower. One glance shows that it's made of florets. A single floret, examined after you cut everything else, looks like a small cauliflower. If you strip that floret of everything except one "floret of a floret"—very soon you must take out your magnifying glass—it's again a cauliflower. A cauliflower shows how an object can be made of many parts, each of which is like a whole but smaller. Many plants are like that. A cloud is made of billows upon billows upon billows that look like clouds. As you come closer to a cloud you don't get something smooth but irregularities at a smaller scale.

Smooth shapes are very rare in the wild but extremely important in the ivory tower and the factory, and besides were my love when I was a young man. Cauliflowers exemplify a second area of great simplicity, that of shapes which appear more or less the same as you look at them up close or from far away, as you zoom in and zoom out.

Before my work, those shapes had no use, hence no word was needed to denote them. My work created such a need and I coined "fractals." I had studied Latin as a youngster and was trying to convey the idea of a broken stone, something irregular and fragmented. Latin is a very concrete language, and my son's Latin dictionary confirmed that a stone that was hit and made irregular and broken up is described in Latin by the adjective "fractus." This adjective made me coin the word "fractal," which now is in every dictionary and encyclopedia. It denotes shapes that are the same from close and far away.

Do I claim that everything that is not smooth is fractal? That fractals suffice to solve every problem of science? Not in the least. What I'm asserting very strongly is that when some real thing is found to be unsmooth, the next mathematical model to try is fractal or multifractal. A complicated phenomenon need not be fractal, but finding that a phenomenon is "not even fractal" is bad news, because so far nobody has invested anywhere near my effort in identifying and creating new techniques valid beyond fractals. Since roughness is everywhere, fractals—although they do not apply to everything—are present everywhere. And very often the same techniques apply in areas that, by every other account except geometric structure, are separate.

To give an example, let me return to the stock market and the weather. It's almost trite to compare them and speak of storms and hurricanes on Wall Street. For a while the market is almost flat, and almost nothing happens. But every so often it hits a little storm, or a hurricane. These are words which practical people use very freely, but one may have viewed them as idle metaphors. It turns out, however, that the techniques I developed for studying turbulence—like weather—also apply to the stock market. Qualitative properties, like the overall behavior of prices, and many quantitative properties as well, can be obtained by using fractals or multifractals at an extraordinarily small cost in assumptions.

This does not mean that the weather and the financial markets have identical causes—absolutely not. When the weather changes and hurricanes hit, nobody believes that the laws of physics have changed. Similarly, I don't believe that when the stock market goes into terrible gyrations its rules have changed. It's the same stock market with the same mechanisms and the same people.

A good side effect of the idea of roughness is that it dissipates the surprise, the irritation, and the unease about the possibility of applying fractal geometry so widely.

The fact that it's not going to lack problems anytime soon is comforting. By way of background, a branch of physics that I

　　　　Benoit Mandelbrot (1924-2010)

was working in for many years has lately become much less active. Many problems have been solved and others are so difficult that nobody knows what to do about them. This means that I do much less physics today than fifteen years ago. By contrast, fractal tools have plenty to do. There's a joke that your hammer will always find nails to hit. I find that perfectly acceptable. The hammer I crafted is the first effective tool for all kinds of roughness and nobody will deny that there is at last some roughness everywhere.

I did not and don't plan any general theory of roughness, because I prefer to work from the bottom up and not from top to bottom. But the problems are there. Again, I didn't try very hard to create a field. But now, long after the fact, I enjoy this enormous unity and emphasize it in every recent publication.

The goal to push the envelope further has brought another amazing development, which could have been described as something recent, but isn't. My book, *The Fractal Geometry of Nature*, reproduced Hokusai's print of the Great Wave, the famous picture with Mt. Fuji in the background, and also mentioned other unrecognized examples of fractality in art and engineering. Initially, I viewed them as amusing but not essential. But I changed my mind as innumerable readers made me aware of something strange. They made me look around and recognize fractals in the works of artists since time immemorial. I now collect such works. An extraordinary amount of arrogance is present in any claim of having been the first in "inventing" something. It's an arrogance that some enjoy and others do not. Now, I reach beyond arrogance when I proclaim that fractals had been pictured forever but their true role remained unrecognized and waited for me to be uncovered.

BOOKS BY JOHN BROCKMAN

UNIVERSE
Leading Scientists Explore the Origin, Mysteries, and Future of the Cosmos

Available in Paperback and eBook

John Brockman brings together the world's best-known physicists and science writers—including Brian Greene, W alter Isaacson, Nobel Prize–winners Murray Gell-Mann and Frank Wilczek, and Brian Cox—to explain the universe in all wondrous splendor.

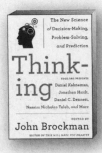

THINKING
The New Science of Decision-Making, Problem-Solving, and Prediction

Available in Paperback and eBook

A cutting-edge exploration of the mysteries of rational thought, decision-making, intuition, morality, willpower, problem-solving, prediction, forecasting, unconscious behavior, and beyond. Edited by John Brockman, publisher of Edge.org, Thinking presents original ideas by today's leading psychologists, neuroscientists, and philosophers who are radically expanding our understanding of human thought.

CULTURE
Leading Scientists Explore Societies, Art, Power, and Technology

Available in Paperback and eBook

This is a cutting-edge master class covering everything you need to know about *Culture*, with original contributions by the world's leading thinkers and scientists. Unparalleled in scope, depth, insight and quality, *Culture* is not to be missed.

THE MIND
Leading Scientists Explore the Brain, Memory, Personality, and Happiness

Available in Paperback and eBook

John Brockman delivers a cutting-edge master class covering everything you need to know about *The Mind*. With original contributions by the world's leading thinkers and scientists, including Steven Pinker, George Lakoff, Philip Zimbardo, V. S. Ramachandran, and others, *The Mind* offers a consciousness-expanding primer on a fundamental topic.

Also Available in Paperback and eBook